Stephan Eggersglüß • Rolf Drechsler

High Quality Test Pattern Generation and Boolean Satisfiability

Stephan Eggersglüß
University of Bremen
Bremen, Germany

German Research Center for Artificial
Intelligence (DFKI) – Cyber-Physical
Systems, Bremen, Germany

Rolf Drechsler
University of Bremen
Bremen, Germany

German Research Center for Artificial
Intelligence (DFKI) – Cyber-Physical
Systems, Bremen, Germany

ISBN 978-1-4419-9975-7 e-ISBN 978-1-4419-9976-4
DOI 10.1007/978-1-4419-9976-4
Springer New York Dordrecht Heidelberg London

Library of Congress Control Number: 2012930124

Printed on acid-free paper

Springer is part of Springer Science+Business Media (www.springer.com)

Preface

The content of the book describes work that has been carried out in the Group of Computer Architecture at the University of Bremen, Germany over the last 5 years. Therefore, we would like to thank all the members of the group for their valuable help. Special thanks go to Daniel Tille and Görschwin Fey for many helpful discussions and support. Both are also co-authors of our previously published book *Test Pattern Generation using Boolean Proof Engines* which describes among other things the basics principles on which the content of this book is based on.

Various chapters of this book are based on scientific papers. Therefore, we would like to acknowledge the work of the co-authors of these papers Görschwin Fey, Hoang M. Le, Juergen Schloeffel and Daniel Tille. Since large parts of the work has been done in collaboration, our special thanks go to the Mentor Graphics Development group in Hamburg, Germany, especially to René Krenz-Bååth (now Hochschule Hamm-Lippstadt). Finally, we would like to thank Lisa Jungmann for her help with the cover design as well as Robert Wille, Judith End and Tom Gmeinder for proof-reading.

Parts of this research work were supported by the German Federal Ministry of Education and Research (BMBF) in the Project MAYA under contract number 01M3172B, by the German Research Foundation (DFG) under contract number DR 287/15-1 and by the Central Research Promotion (ZF) of the University of Bremen under contract number 03/107/05. The authors like to thank these institutions for their support.

Bremen

Stephan Eggersglüß
Rolf Drechsler

Contents

List of Figures

List of Tables

List of Acronyms

ARAP	As-Robust-As-Possible
ATE	Automatic Test Equipment
ATPG	Automatic Test Pattern Generation
BCP	Boolean Constraint Propagation
BDD	Binary Decision Diagram
CNF	Conjunctive Normal Form
CUT	Circuit Under Test
DCA	Dynamic Clause Activation
DFT	Design-For-Test
DTPG	Deterministic Test Pattern Generation
GDFM	Gate Delay Fault Model
IC	Integrated Circuit
IG	Implication Graph
ISAT	Incremental SAT
LC	Logic Class
LWL	Learned Watch List
ODC	Observability Don't Care
PBO	Pseudo-Boolean Optimization
PB-SAT	Pseudo-Boolean SAT
PDF	Path Delay Fault
PDFM	Path Delay Fault Model
PI	Primary Input
PO	Primary Output
PPI	Pseudo Primary Input
PPO	Pseudo Primary Output
RTPG	Random Test Pattern Generation
s-a-0	Stuck-at-0
s-a-1	Stuck-at-1
SAFM	Stuck-At Fault Model
SAT	Boolean Satisfiability, Satisfiable

SDD	Small Delay Defect
SWL	Structural Watch List
TF	Transition Fault
TFM	Transition Fault Model
UNSAT	Unsatisfiable

List of Symbols

$\downarrow$	falling transition
$\uparrow$	rising transition
$\cdot$	Boolean AND operator
$+$	Boolean OR operator
$\odot$	resolution operator
$\oplus$	Boolean XOR operator
$\bar{\cdot}$	Boolean NOT of $\cdot$
$\rightarrow$	implies
$\leftrightarrow$	equivalence
Φ	CNF, set of clauses
Φ_{dyn}	dynamically extended CNF
Φ_F	fault-specific constraints
Ψ	set of pseudo-Boolean constraints
ψ	pseudo-Boolean constraint of Ψ
η	Boolean encoding
κ	conflict
λ	arbitrary literal
ω	clause of Φ
ω_C	conflict clause
$@x$	at decision level x
$\mathbb{B}$	set of Boolean values $\{0,1\}$
$\mathscr{C}$	circuit
$\mathscr{F}$	set of flip-flops
F	fault
f	flip-flop $\in \mathscr{C}$, faulty line of gate
$\mathscr{F}(g)$	transitive fanin of g
$\mathscr{G}$	set of gates
g	gate $\in \mathscr{C}$
h	successor gate of g
$\mathscr{I}$	set of primary inputs
$i_1,\ldots,i_n$	primary inputs of $\mathscr{C}$

$\mathscr{J}$	J-stack
$\mathscr{L}_x$	multiple-valued logic with x values
$\mathscr{O}$	set of primary outputs
$o_1, \ldots, o_m$	primary outputs of $\mathscr{C}$
$\mathscr{P}$	structural path of $\mathscr{C}$
$\mathscr{S}$	set of signal lines or connections
s	signal line, connection
t_i	initial (current) time frame
t_{i+1}	final (next) time frame
V	vector
v	(Boolean) value
X	set of Boolean variables, don't care value
$x_1, \ldots, x_n$	Boolean variables

Chapter 1
Introduction

The *Integrated Circuit* (IC) was invented in the 1950s. At the beginning, ICs were mainly used in computers. With the advancing miniaturization of the components, the significance of ICs as part of our daily life grows. Many consumer products such as mobile music players or cell phones use ICs (or "chips") as core engines. A failure of these devices usually results in problems of lesser extent for the owner. But today, ICs also control safety critical applications. For example, chips are responsible for the correct mode of operation in car control systems, avionics or medical equipment. A failure of a chip can be life-threatening in the worst-case. Consequently, the correct mode of operation of a fabricated chip is crucial.

Due to the ever shrinking component sizes of today's designs, the vulnerability of chips to flaws in the manufacturing process increases. The IC manufacturers put much effort in guaranteeing the integrity of their products. A large part of the manufacturing costs is spent for the detection of defects caused by the manufacturing process. Every fabricated chip is subjected to a *post-production test* (or *manufacturing test*) to avoid that defective chips are delivered to customers (and by this could cause failures in operation mode). Thus, the purpose of such a post-production test is to detect any defects caused by the manufacturing process.

Stimuli are applied to the inputs of the *Circuit Under Test* (CUT) during this test. The output responses are monitored. If one or more of the output responses are inconsistent with the specification, the chip will be rejected as erroneous. However, the complexity of modern designs does not allow for a complete test of all possible input stimuli. A complete test for each fabricated chip would be far too time-consuming or costly, since the number of possible tests is exponential in the number of inputs. Instead, a test set is pre-computed that covers a large range of possible defects. Logical fault models are used to abstract from physical defects. The fault model most widespread is the *Stuck-at Fault Model* (SAFM) [Eld59].

This test set is applied to each fabricated chip by *Automatic Test Equipment* (ATE). Because the memory and bandwidth of an ATE is limited, the applied

S. Eggersglüß and R. Drechsler, *High Quality Test Pattern Generation and Boolean Satisfiability*, DOI 10.1007/978-1-4419-9976-4_1,

test set has to be as small as possible. A large test set size signifies not only a long test application time but also immense test costs. The computation of the test set, which is known as *Automatic Test Pattern Generation* (ATPG), is the main subject of this book. Due to the large number of potential faults, ATPG is a computationally intensive task and fast algorithms are needed for obtaining a test set in acceptable run time.

Classical ATPG algorithms are mostly based on the D-algorithm proposed by Roth in 1966 [Rot66]. These algorithms work directly on the circuit structure, i.e. a flat gate-level netlist, and typically benefit significantly from their knowledge about the structure of the problem. Several techniques and powerful heuristics have been proposed over the years to improve the effectiveness of ATPG. At the turn of the millennium, the ATPG problem was considered to be solved. The existing algorithms were fast enough and provided sufficient fault coverage. This has changed in the last years.

The size of new designs doubles every 18 months according to Moore's law [Moo65]. Today's circuits consist of multi-million gates and the classical structural ATPG algorithms reach their limits. Tests for a large number of faults can still be generated very quickly. But the size of the set of faults for which no test can be generated in acceptable run time increases significantly. As a result, the high fault coverage demands of the chip manufacturers can barely be met. Thus, the overall quality of the test set decreases due to the lower fault coverage. As a consequence, there is a need for new robust ATPG algorithms especially when considering future design sizes.

Furthermore, another crucial point has been emerged in the field of manufacturing test. Due to the increased operation speed beyond the GHz mark and the small manufacturing technologies, the number of timing-related defects which affects the product quality has increased. This trend is expected to grow with the process technology scaling down towards very deep sub-micron devices. Therefore, *delay testing* has become mandatory to filter out defective devices and to assure that the desired performance specifications are met.

Test generation for delay faults generally needs more computational effort than ATPG for stuck-at faults because test patterns have to be computed over at least two time frames. A large number of delay faults remain unclassified during ATPG for modern designs. Moreover, in contrast to the SAFM, test patterns for delay faults can be classified in different quality levels like robust and non-robust test patterns [KC98]. The quality of a test can be – simply spoken – defined as the probability of fault detection. High quality test patterns are more desirable but usually harder to obtain. As a result, the need for new robust ATPG algorithms is even more urgent in the field of delay test generation. In the last years, *Small Delay Defects* (SDDs) have become more and more serious. Therefore, the ability to detect SDDs has become an important indicator for judging the quality of a delay test. However, ATPG approaches struggle with the generation of tests dedicated to detect SDDs due to the high complexity.

A promising solution to close the gap between test quality requirements and ATPG effectiveness is the application of solvers for *Boolean Satisfiability* (SAT).

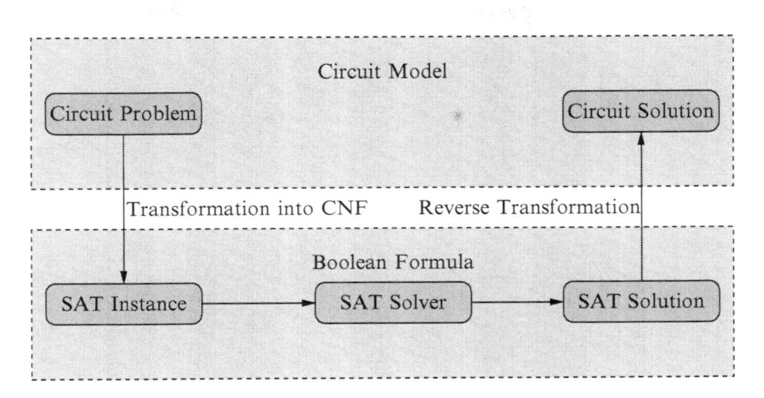

Fig. 1.1 SAT application flow

The SAT problem was the first problem proven to be NP-complete by Cook in 1971 [Coo71]. Unlike classical ATPG algorithms, SAT solvers do not work on a circuit representation but on a Boolean formula f in *Conjunctive Normal Form* (CNF). The SAT problem is defined as the question whether there exists an assignment a such that the Boolean formula $f(a)$ evaluates to 1 and if the problem is satisfiable, to construct a satisfying model.

Since the 1990s, powerful SAT techniques have been developed and the efficiency of SAT solvers is still increasing. The homogeneity of the underlying CNF permits the application of efficient implication techniques and powerful conflict analysis strategies to solve SAT problems. The utilization in the field of design verification and other problem domains has shown that SAT-based algorithms are very robust for large and hard problem instances. The general flow for applying a SAT solver to a circuit-oriented problem, e.g. ATPG, is shown in Fig. 1.1. The original problem which is based on a circuit model must be transformed into a SAT instance in CNF. Then, the SAT solver is applied to the CNF to solve the formula. Finally, the obtained solution must be transformed from the SAT model to the original circuit model. However, it has to be noted that the SAT solver itself works as a black-box and does not know anything about the origin of the problem. The solving process is typically completely independent from the SAT formulation.

The efficiency of state-of-the-art SAT solvers and the advances in SAT techniques have led to an incorporation of SAT techniques in other problem domains as well. For instance, solvers for *Pseudo-Boolean Optimization* (PBO) often rely strongly on these efficient SAT techniques.

Recently, the SAT-based ATPG tool PASSAT [SFD$^+$05, DEF$^+$08] has been proposed for stuck-at test pattern generation in an industrial environment. The first results of PASSAT for industrial circuits were very promising. In particular, many hard-to-detect faults for which classical ATPG algorithms failed to generate a test could be solved by the SAT-based algorithm. The results further show that SAT-based algorithms have the potential to close or significantly reduce the gap between test quality requirements and ATPG effectiveness. However, PASSAT as one of the most advanced SAT-based ATPG approaches has several shortcomings that prevent the efficient use in industrial practice. The main shortcomings are listed in the following:

- Run time – SAT-based ATPG shows promising results in classifying hard-to-detect faults. However, there is overhead for easy-to-detect faults which typically represent the majority of faults. As a result, the overall run time is often not acceptable.
- Test pattern compactness – The ATPG process in an industrial environment is part of a greater flow. In order to obtain a small test set, test patterns are subsequently processed by techniques like test compaction or test compression. Typically, SAT-based ATPG generates over-specified test patterns which usually increases the size of the test set significantly. This is not acceptable since it directly results in higher test costs.
- Delay fault models – Test generation in an industrial environment is typically done for more than one fault model. Often, SAT-based ATPG approaches were developed specially for the SAFM. For instance, the initial PASSAT approach does not consider other fault models than the SAFM, in particular not any delay fault model and the quality aspect.
- Small delay defects – The detection of faults caused by SDDs has become an important issue in the manufacturing test. The generation of particular tests to detect these faults requires the processing of timing information as well as other constraints like the detection via the longest path. Since the SAT problem is a Boolean decision problem and not an optimization problem, classical SAT-based algorithms are not suited for these kind of problems.

The aim of the techniques and methodologies presented in this book is to improve SAT-based ATPG and address the shortcomings listed above in order to make SAT-based ATPG applicable in industrial practice. The contributions improve the performance and robustness of the overall test generation process. Here, the term *robustness* means that the ATPG algorithm should reliably generate test patterns for most targeted faults in acceptable run time to meet the high fault coverage demands of the industry. Another focus was to increase the quality of the test set. The techniques and improvements presented in this book provide the following advantages:

- Advances in SAT solving techniques – Circuit-oriented SAT solving techniques are introduced. These techniques make use of structural information and, thus, they are able to accelerate the search process significantly. Although these solving techniques have been developed for ATPG, they are applicable to other circuit-oriented problems as well.
- New SAT formulations for delay fault models – Besides the classical stuck-at fault model, SAT formulations for the prevalent delay faults models are provided since these fault models become more and more important.
- Test quality improvements – Special attention is paid to the efficient generation of high-quality tests. These tests are more desirable but typically harder to obtain.

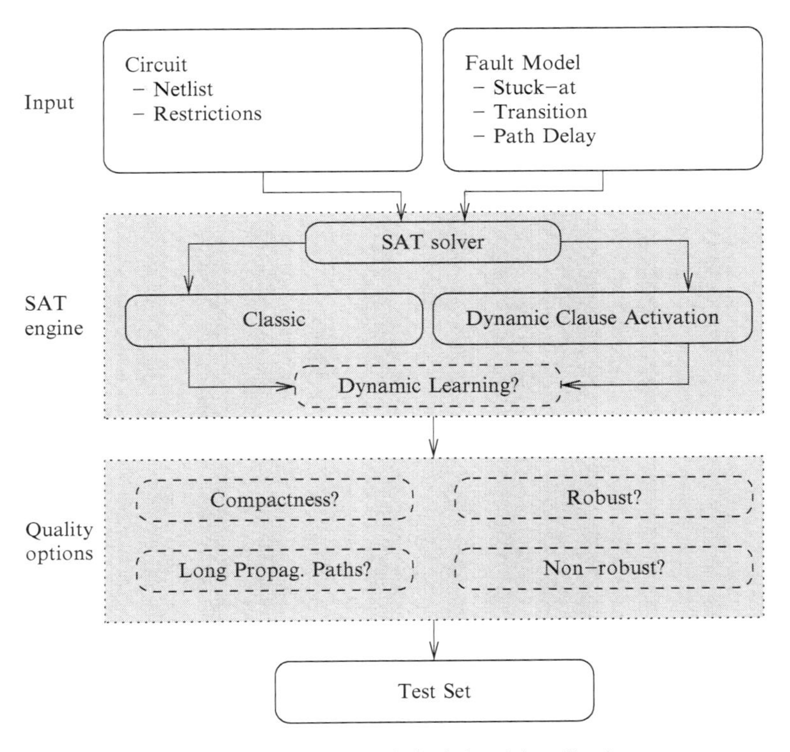

Fig. 1.2 Developed SAT-based ATPG framework for industrial application

Altogether, a SAT-based ATPG framework well suited for the use in an industrial test environment results. Figure 1.2 shows an illustration of this framework. The framework processes industrial circuits and can be applied for the fault models most widely used in industrial practice as shown in the upper part of the illustration. The core engine of the framework is a SAT solver which is enhanced with the techniques proposed in this book in order to improve the performance and robustness. Additionally, certain options can be used to improve the quality or compactness of the test patterns. The framework is able to reduce the gap between the test quality requirements of the industry and the ATPG effectiveness significantly.

The proposed techniques were implemented and integrated into the ATPG framework of NXP Semiconductors as a prototype in order to show the benefits which can be obtained using SAT-based ATPG in industrial practice. Experiments on large industrial circuits with several million elements showed the robustness and feasibility of the proposed techniques.

In addition, the field of application was extended to other problem domains as well. Since many current serious problems in test generation, e.g. test pattern generation for SDDs, cannot directly be formulated as a Boolean SAT problem, related fields are explored. For instance, the development for PBO solvers has been largely influenced by the advances in the SAT domain and new SAT techniques can often be transferred to this domain. This book shows how the problem formulation must be modified in order to leverage the benefits of a PBO solver. In particular, the two following techniques, which demonstrate the potential of PBO-based algorithms, are presented:

- PBO formulation for timing-aware ATPG [LTW+06] – In order to make use of the efficient SAT/PBO solving techniques, a PBO problem formulation for timing-aware ATPG is given.
- New quality level – Since there is a large quality gap between non-robust and robust tests, the generation of *As-Robust-As-Possible* (ARAP) tests using a PBO-based algorithm is shown in order close the quality gap as far as possible.

The book is divided into three parts. Please note that some techniques presented in this book are based on research work presented in a previous book [DEFT09]. Parts of the preliminaries are taken from [DEFT09] and the relevant techniques are briefly reviewed in Part to make this book self-contained.

- Part I presents the preliminaries to understand the research work as well as an overview of the previous work done in this field. This part contains basic information about testing in general, delay fault models, classical structural ATPG approaches as well as about SAT and algorithms to solve the SAT problem. Additionally, the initial SAT-based ATPG approach PASSAT and early extensions as discussed in [DEFT09] are briefly reviewed.
- Part II presents the developed advances in SAT solving techniques boosting the performance and robustness especially of SAT-based ATPG. The proposed techniques leverage the characteristics of the ATPG problem in order to achieve significant improvements compared to the initial approach. By this, the advantages of structural and SAT-based ATPG can be combined. The generated SAT instances are often very similar, because many SAT instances stem from the same circuit part. This fact is exploited in particular in the discussed techniques.
- Part III deals with the application of SAT-based algorithms to delay fault models. In particular, high quality test generation is focused in this part. SAT formulations for the delay fault models most widely used in industrial practice, i.e. the *Transition Fault Model* (TFM) and the *Path Delay Fault Model* (PDFM), are proposed in order to allow for a wide application of SAT-based ATPG in an industrial test environment. This part also shows how PBO solvers can be used to increase the quality of delay tests.

A brief presentation of the content of each remaining chapter is given to conclude the introduction:

Part I: Preliminaries and Previous Work

- Chapter 2 presents elementary information about the role of ATPG in the production test as well as the abstraction level of the circuits used and the usage of fault models, especially delay fault models. Furthermore, the basic concepts

of classical ATPG algorithms as well as a sketch of the general structure of an industrial ATPG environment are given.
- Chapter 3 describes the basics of SAT applications. State-of-the-art SAT solving techniques like for instance conflict analysis, the general circuit-to-CNF transformation procedure as well as the usage of observability don't cares are explained.
- Chapter 4 shows the basic problem formulation of SAT-based ATPG for the SAFM as used in the initial PASSAT approach. The usage of a multiple-valued logic and a Boolean encoding, respectively, are explained. Furthermore, a combination of a SAT-based ATPG algorithm and a classical structural algorithm to exploit the advantages of both is presented.

Part II: New SAT Techniques and Their Application in ATPG

- Chapter 5 presents the SAT technique *Dynamic Clause Activation* to speed up the solving process for circuit-oriented problems. For this, a data structure called *Structural Watch List* is proposed and an efficient activation methodology is provided. Furthermore, the emulation of a classical SAT solver is presented which increases the flexibility of the SAT framework. Moreover, it is shown how this technique can be applied to multiple-valued logic.
- Chapter 6 proposes a dynamic learning scheme for circuit-oriented problems. The efficient passing of learned information from one circuit-based SAT instance to another is presented. Additionally, different strategies to identify and discard "useless" information are given. A brief overview of the improved solving engine which incorporates the techniques of the chapter as well as of the previous chapter is given.

Part III: High Quality Delay Test Generation

- Chapter 7 describes SAT-based ATPG for the TFM. The usage of an *Iterative Logic Array* to represent two time frames is explained. The SAT modeling of transition faults by injecting stuck-at faults is presented. Experimental results on large industrial circuits show that the fault coverage can be raised by up to 2% using SAT-based ATPG. Additionally, techniques to prioritize long propagation paths are proposed to increase the quality level of the generated test patterns. Moreover, a PBO formulation for timing-aware ATPG is given which is able to process timing information and generate high-quality tests which detects faults through the longest path using transition-dependent delays.
- Chapter 8 presents techniques for SAT-based ATPG for the PDFM. In order to generate robust test patterns in an industrial environment, a multiple-valued logic has to be used which results in excessive large SAT instances. A SAT instance

generation flow is shown which uses structural properties of the circuit to shrink the size of the resulting SAT instances by using several multiple-valued logics with fewer values. Furthermore, a PBO-based ATPG algorithm for generating *As-Robust-As-Possible* tests is introduced.
- Chapter 9 summarizes the contributions and gives conclusions. Furthermore, possible directions of future research are pinpointed.

Part I
Preliminaries and Previous Work

Chapter 2
Circuits and Testing

This chapter gives the basic information about circuits and testing of circuits. The role of *Automatic Test Pattern Generation* (ATPG) in the production test is presented in Sect. 2.1. Section 2.2 gives information about the used abstraction level of circuits and shows the modeling as well as the basic notations for the circuit representation used. In Sect. 2.3, the meaning of a fault model is described and relevant fault models are introduced, while classical algorithms for test pattern generation for these fault models are presented in Sect. 2.4. Section 2.5 briefly reviews the industrial test environment.

2.1 Post-Production Test

The manufacturing process of a circuit is very vulnerable to defects of different kinds. As a matter of fact, defects created during the manufacturing process can not be avoided, especially due to shrinking feature sizes. However, the delivery of defective chips to customers has to be prevented. Therefore, each manufactured chip is tested for its functional correctness by a post-production test.[1] The test of a circuit is depicted in Fig. 2.1. Input stimuli are applied consecutively at the *Primary Inputs* (PIs) of the *Circuit Under Test* (CUT) and the responses are monitored at the *Primary Outputs* (POs). If a received response differs from the expected one, the chip is classified as erroneous. Input stimuli applied in order to test the correct function of the CUT are also referred to as tests or test patterns. A test vector is referred to as a single assignment of the PIs. In contrast, a test or test pattern can be a single test vector or multiple test vectors applied consecutively.

However, a complete test of all possible input stimuli is not feasible. The number of possible input stimuli is exponential in the number of inputs. For example, a chip

[1]Note that it is basically assumed for this test that the design itself is free of errors.

S. Eggersglüß and R. Drechsler, *High Quality Test Pattern Generation and Boolean Satisfiability*, DOI 10.1007/978-1-4419-9976-4_2,

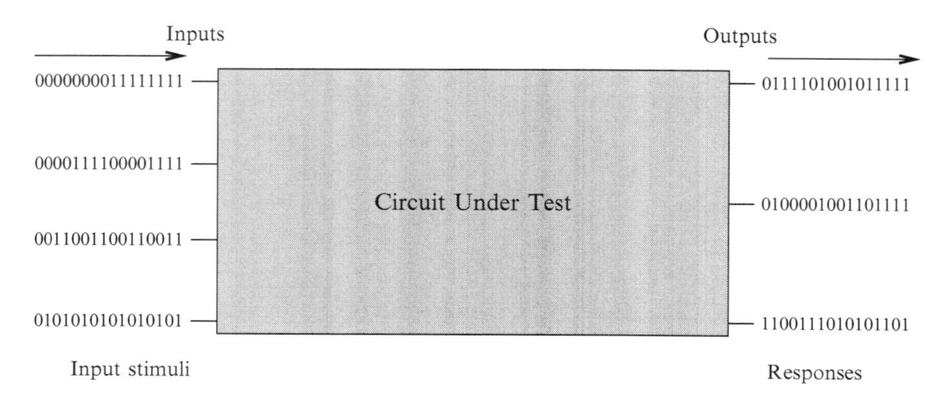

Fig. 2.1 Circuit under test

with 64 inputs has $2^{64} = 18{,}446{,}744{,}073{,}709{,}551{,}616$ possible test vectors. This is not acceptable with respect to the test time as well as to the test costs. In order to overcome this problem, a subset of all possible input stimuli has to be chosen, i.e. the test set. This test set should be small at the one hand but on the other hand be able to detect a large number of possible defects.

Crucial for the test set computation is the use of *fault models*. A fault model is a mathematical abstraction of a physical defect and describes the logic behavior of the defect. The use of fault models allows the application of efficient algorithms. Fault models are essential for an efficient test set computation due to the large number of possible physical irregularities. The fault model most widely used in practice is the *Stuck-At Fault Model* (SAFM). Here, a signal line (or connection) in the circuit is permanently "stuck" at a specified value, i.e. 0 or 1. The SAFM is well-understood and known to detect a wide variety of physical defects [JG03]. In an optimal test set, each fault of a particular fault model, e.g. a stuck-at-0 and a stuck-at-1 fault for each connection in the CUT, is detected by at least one test pattern contained in the test set or is known as untestable.

The computation of the test set is generally known as ATPG. The ATPG process involves complex calculations and is typically performed only once for each design. The generated test set is applied to each manufactured chip. Different ATPG techniques can be used for the computation of the test patterns. *Random Test Pattern Generation* (RTPG) randomly generates input stimuli for the CUT, i.e. a test pattern. A fault simulator is used in order to detect which faults are detected by this test pattern. RTPG is fast, but suffers from the circumstance that, in most cases, only a low fault coverage can be reached in reasonable time. Furthermore, untestability cannot be proven. A high fault coverage is mandatory to maintain a certain quality level for chips delivered to customers.

In contrast, *Deterministic Test Pattern Generation* (DTPG) takes a particular fault as input and generates a test pattern which detects this fault explicitly or proves that no test pattern for this fault exists. By this, high fault coverage can be achieved because each single fault is targeted. However, DTPG is more complex

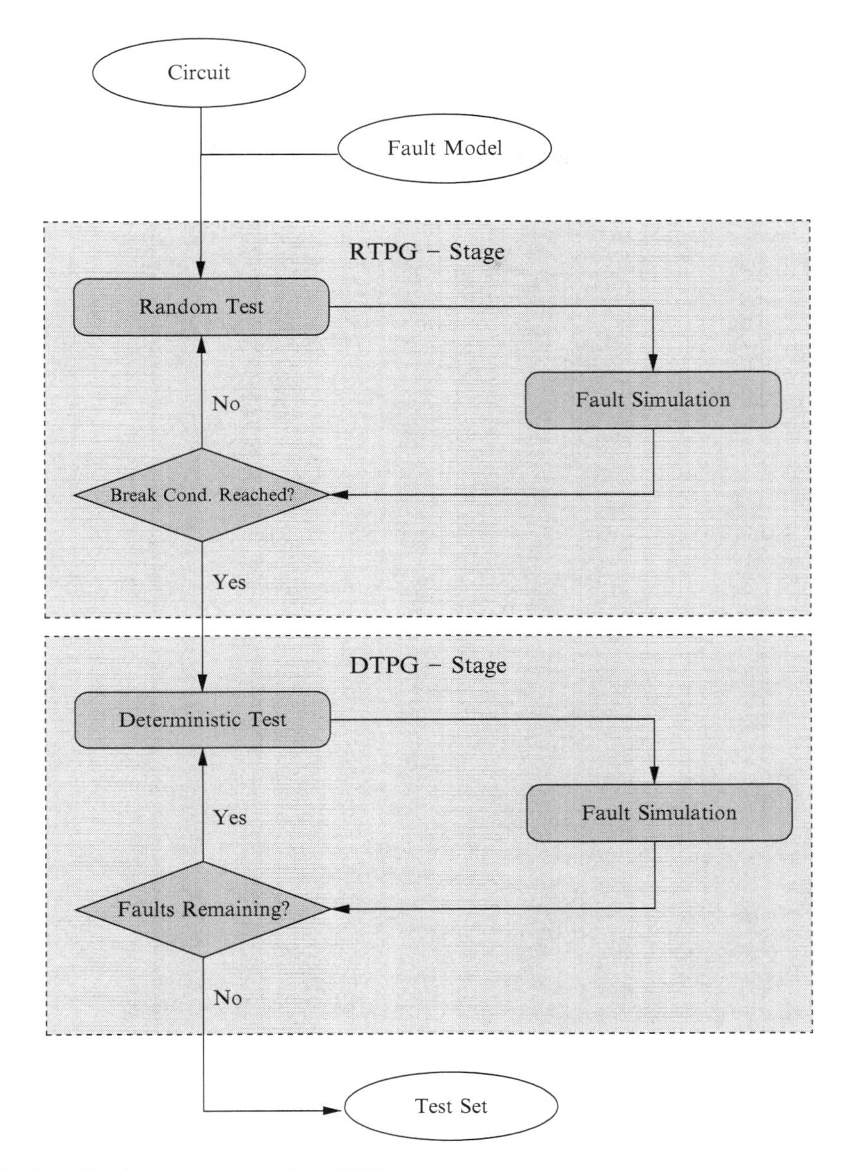

Fig. 2.2 Simple representation of an ATPG system

than RTPG and consequently needs more run time for a single call. In fact, DTPG for a particular fault is proven to be an NP-complete problem for arbitrary digital circuits [FT82]. In modern ATPG systems, a combination of fault simulation, RTPG and DTPG is employed resulting in a good trade-off between quality and run time. Figure 2.2 shows a simplified version of such an ATPG system.

First, RTPG is started as a pre-process to generate test patterns for the easy-to-detect faults. By this, a large number of faults does not have to be targeted by DTPG. For the remaining faults, DTPG is invoked. The fault simulator is used to compute additionally detected faults for each generated test pattern. DTPG has not to be executed anymore for these faults, since there already exists a test pattern that detects them. If all faults are either untestable or detected by a generated test, the ATPG process is finished. The most time consuming task of this procedure is the DTPG-stage. The improvement of DTPG is topic of this book. In the following, DTPG is generally referred to as ATPG.

2.2 Circuits

A circuit $\mathscr{C}$ can be represented in different abstraction levels, for instance in *register-transfer level, gate level, switch level* or *physical level* [WWW06]. Circuits are usually modeled in gate level representation for test generation.

Definition 2.1. A combinational circuit in gate level representation is a directed acyclic graph $\mathscr{C} = (\mathscr{G}, \mathscr{S}, \mathscr{I}, \mathscr{O})$, where

- $\mathscr{G}$ is the set of gates,
- $\mathscr{S}$ is the set of signal lines or connections,
- $\mathscr{I}$ is the set of primary inputs and
- $\mathscr{O}$ is the set of primary outputs.

Gates are usually denoted by lower case latin letters. Here, additional indices can be used for ordering. A gate g of circuit $\mathscr{C}$ is often denoted by $g \in \mathscr{C}$. The underlying graph structure of $\mathscr{C}$ is created by signal lines between gates. A signal line s connects exactly two gates g, h to each other (denoted by $s = g \times h$). Gate g is the *predecessor* of h, while h is referred to as *successor* of g. Signal lines are usually denoted by the notation of the predecessor gate, i.e. g. Additionally, if variables are assigned to a signal line, the same notation is used.

Each gate has a specific type defining the function of the gate. The basic gates or gate types used in this book are depicted in Fig. 2.3. The gates INV (NOT), AND, OR and XOR correspond to the logical operators (see also Sect. 3.1). The gates NAND, NOR and XNOR (EQUIVALENCE) correspond to the inverted versions of these operators. The gate BUF represents the identity function.

More complex primitive gates are the multiplexer (MUX) and the BUS or busdriver (BUSDRV), respectively. Multiplexer and busdriver are non-symmetric gates. Therefore, a unique order of the inputs, i.e. the predecessors, is necessary. Additionally, BUS and BUSDRV are not Boolean gates but tri-state elements with an additional state of high impedance. All gate types presented until now have only one successor. FANOUT gates can have multiple successors and are employed to represent branches in $\mathscr{C}$. In particular, FANOUT gates are important for fault modeling. The outgoing branches of a FANOUT gate are typically not denoted by the predecessor gate, but have own notations.

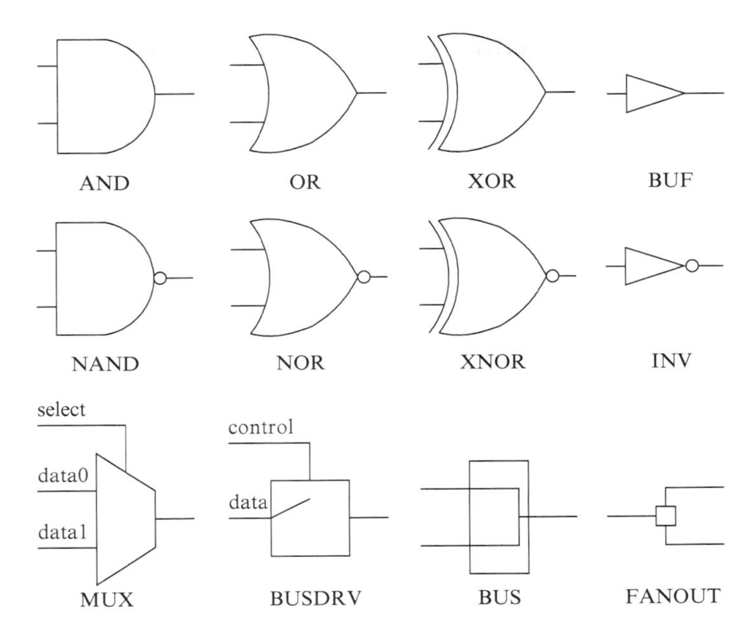

Fig. 2.3 Basic gates

A gate with no incoming connections is a PI, while a gate with no outgoing connection is a PO of $\mathscr{C}$. The PIs of $\mathscr{C}$ are often denoted by $i_1, \ldots, i_n$, while the POs are often denoted by $o_1, \ldots, o_m$. The *transitive fanin* of a gate g is the set of gates on which g is structurally dependant. This is denoted by $\mathscr{F}(g)$.

In contrast to a combinational circuit, a sequential circuit $\mathscr{C}$ contains state elements or flip-flops. A sequential circuit is a circuit whose output value depends not only on the applied input values but also on the current state of the circuit, i.e. values stored in the state elements. The next state of the circuit is likewise computed by the combinational part, the input values and the current state.

Definition 2.2. A sequential circuit is described as $\mathscr{C} = (\mathscr{G}, \mathscr{S}, \mathscr{I}, \mathscr{O}, \mathscr{F})$ where $\mathscr{G}, \mathscr{S}, \mathscr{I}, \mathscr{O}$ denote the elements of a combinational circuit as given in Definition 2.1 and $\mathscr{F}$ is the set of flip-flops contained in $\mathscr{C}$.

Flip-flops store the result of the combinational part of the circuit in time frame t_i and propagate it in the next time frame t_{i+1}. Here, a time frame is specified as the duration of a clock period. Generally, this is referred to as *sequential behavior*. A schematic view of a sequential circuit is shown in Fig. 2.4.

2.2.1 Scan-Based Testing

Scan-based testing – in contrast to functional testing – is used to decrease the complexity of the test of sequential circuits. The use of storage elements like

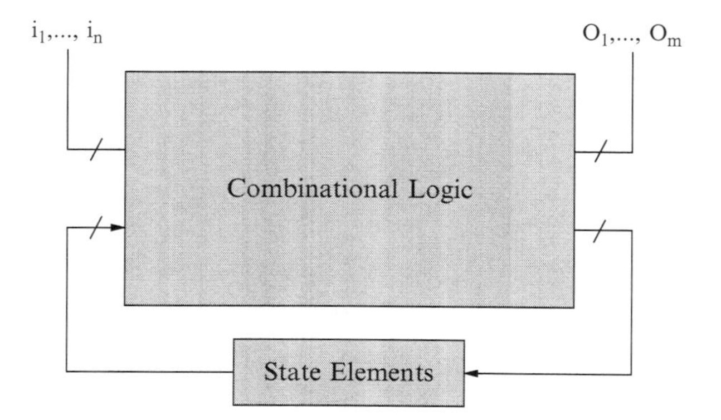

Fig. 2.4 Sequential circuit

flip-flops requires pre-initialization in order to detect certain faults. This means, the storage elements must be loaded with the desired values by the application of a sequence of vectors. This imposes overhead for the test generation as well as to the test application. In contrast, scan techniques are used to reduce this overhead. Using scan-based testing, the flip-flops are connected in several *scan-chains* [WA73, EW77]. Due to the use of scan chains, each flip-flop $f \in \mathscr{F}$ can be pre-initialized with an arbitrary value during a shift-mode before the application of the test. After the test has been applied, the response value of the combinational logic $\mathscr{F}(f)$ can be shifted out, i.e. an observation is possible. As a consequence, each flip-flop $f \in \mathscr{F}$ is splitted into a *Pseudo Primary Input* (PPI) – output of f – and a *Pseudo Primary Output* (PPO) – input of f. That means, for a one-vector-test, each PPI (PPO) can be treated as a PI (PO).

For two-vector-tests which are usually applied in delay testing, different scan modes were introduced. The most prevalent scan modes are *enhanced-scan* [DS91], *launch-on-shift (skewed-load)* [SP93] and *launch-on-capture (broad-side)* [SP94].

In the enhanced-scan mode, no relation between both patterns is required. An arbitrary vector-pair can be applied which makes the test generation easier. However, this requires a special scan-design (hold-scan) which causes overhead. In a skewed-load scan mode, the second test vector is a shifted version of the first vector. During broad-side testing, the second vector is the functional response of the first vector. Throughout this book, the broad-side testing technique is assumed. Tests for this scheme are more complex to compute since the logical behavior of two time frames has to be considered.

2.3 Fault Models

During the manufacturing process, a large range of physical defects may enter the design causing malfunctions. Chips with malfunctions have to be filtered out by the post-production test before being delivered to customers. However, testing for each single known defect is impractical because of the large number of potential defects and the associated excessive computational effort. Therefore, "logical" fault models are introduced. A fault model is a mathematical abstraction and models the logical behavior of physical defects.

The use of a fault model has several advantages [ABF90]. Different physical defects can be modeled by a single fault model. Even physical defects which are not fully understood can be covered. Additionally, fault models are technology-independent. The developed test generation methods do not have to be modified when the underlying technology changes. Furthermore, the complexity of test generation is significantly reduced due to the logic modeling. Therefore, fault models are essential for an efficient test set computation. Relevant fault models are introduced in this section.

First, the basic *Stuck-at Fault Model* (SAFM) is introduced in detail in Sect. 2.3.1. Then, the most common delay fault models are described in Sect. 2.3.2. For a detailed overview on a large range of existing fault models, it is referred to [BA00].

2.3.1 Stuck-at

The SAFM [Eld59, GNR61] is well-understood and the fault model most widely used in practice. Although being very simple, the SAFM is known to detect a large number of potential defects. The SAFM belongs to the group of static fault models. Static fault models affect the logic function of the circuit.[2] A signal line f (short: line) is assumed to be "stuck" at a fixed value and does not depend on the input values or on $\mathscr{F}(f)$, respectively, anymore. There are two different faults associated to each line. When the line is stuck at the value 0, it is called *stuck-at-0* (s-a-0) fault. Otherwise, when the line is stuck at the value 1, the fault is called *stuck-at-1* (s-a-1) fault. The number of stuck-at faults in a circuit $\mathscr{C}$ is linear in the number of signal lines, i.e. for n signal lines in $\mathscr{C}$, there are $2n$ possible stuck-at faults. Note that for FANOUT gates, each outgoing line counts as possible fault location for which a test has to be generated. Formally, a stuck-at fault F is denoted by a tuple (f, v) where f is the faulty signal line and v is the fault value, i.e. 0 for a s-a-0 fault and 1 for a s-a-1 fault. This is shown in the following example.

Example 2.1. Consider Fig. 2.5 where a simple example circuit is presented. Figure 2.5a shows the correct circuit, while in Fig. 2.5b, c, s-a-0 faults are injected on each outgoing "branch" of the FANOUT gate, respectively. By injecting the fault, i.e. a constant value, the branch is disconnected from the AND gate a.

The *Multiple SAFM* assumes the presence of multiple stuck-at faults. However, this fault model is rarely used in practice due to the huge number of faults.

[2] Other static fault models are e.g. the bridging fault model or the cellular fault model. In this book, we consider only the SAFM. The results can be transferred.

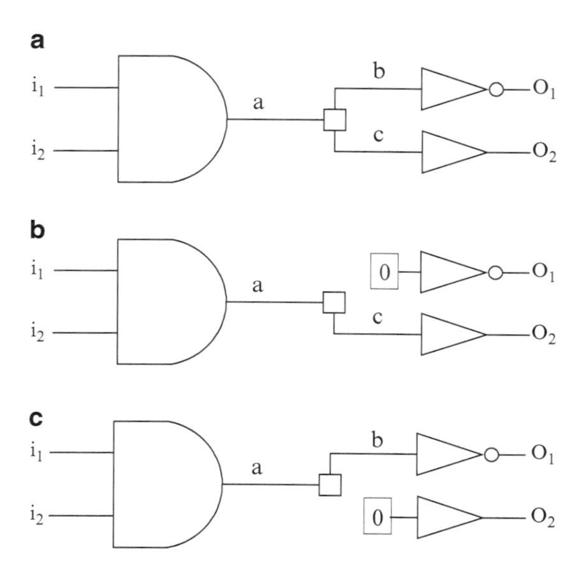

Fig. 2.5 SAFM example. (**a**) Correct circuit, (**b**) stuck-at-0 fault on branch *b* and (**c**) stuck-at-0 fault on branch *c*

2.3.1.1 Test Generation

A test pattern or simply "test" for a stuck-at fault consists of one input vector V which activates the fault at the fault site and propagates the fault effect to an observation point, i.e. a PO or PPO. The chosen propagation path has to be logically sensitized for the propagation of the fault effect. A method for creating test patterns is the computation of the *Boolean Difference* [SHB68]. Here, the correct and the faulty circuit form a circuit similar to a *miter* circuit [Bra93], as it can be used for combinational equivalence checking. The inputs of both the correct and faulty circuit are connected in a miter to ensure that both circuits assume the same input values. The outputs of both circuits are compared using XOR gates. The results of the comparisons serve as inputs of an OR gate. Finally, the output of the OR gate is constrained to the value 1. By this, it is ensured that at least one original output produces a different result in the correct and faulty circuit.[3] The following example demonstrates the modeling.

Example 2.2. Consider the correct circuit given in Fig. 2.5a and the circuit with an injected stuck-at fault $(b,0)$ presented in Fig. 2.5b. The resulting miter circuit is shown in Fig. 2.6. In this simple example, four different input vectors exist.

[3]For reason of simplicity, the method is described by using gates. Boolean expressions are used instead of gates in the original work.

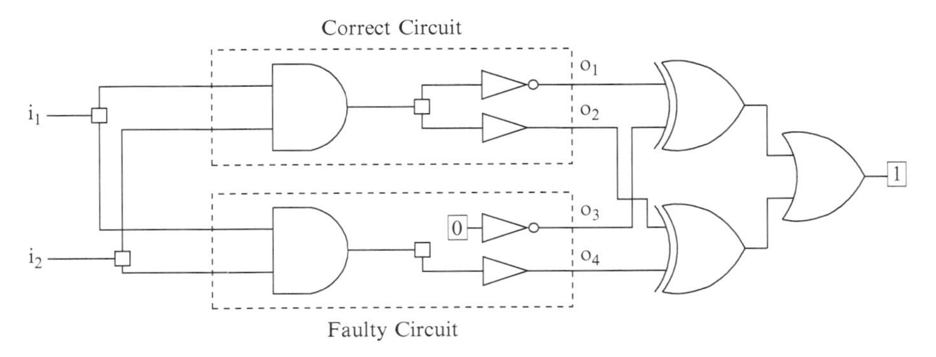

Fig. 2.6 Boolean difference

Table 2.1 Input vectors

Vector	Inputs i_1	i_2	Outputs o_1	o_2	o_3	o_4	OR
V_1	0	0	1	0	1	0	0
V_2	0	1	1	0	1	0	0
V_3	1	0	1	0	1	0	0
V_4	1	1	0	1	**1**	1	**1**

Table 2.1 shows the input vectors and the resulting output values. Only input vector V_4 produces a difference on the output. Therefore, the input vector V_4 is a test pattern for the stuck-at fault $(b,0)$.

Faults for which at least one test pattern exists are called *testable*. If no test pattern exists, the fault is called *untestable* or *redundant*. The question whether a fault is testable or untestable is an NP-complete problem [FT82]. More elaborated and more efficient ATPG algorithms are described in Sect. 2.4.

2.3.1.2 Fault Collapsing

The number of faults which have to be considered by the ATPG process is typically very huge. Therefore, *fault dominance* and *fault equivalence* relationships are exploited in order to reduce the number of faults, which have to be processed by ATPG.

Definition 2.3. A stuck-at fault F_1 *dominates* a fault F_2 if every test pattern that detects F_2 also detects F_1. When F_1 dominates F_2 and F_2 dominates F_1 as well, F_1 and F_2 are said to be *equivalent*.

If F_1 dominates F_2, a test has to be generated only for F_2. If F_1 and F_2 are equivalent, a test can be generated either for F_1 or F_2 to detect both faults. The procedure to reduce the target fault set by exploiting fault dominance relationships is called *fault collapsing* [ABF90].

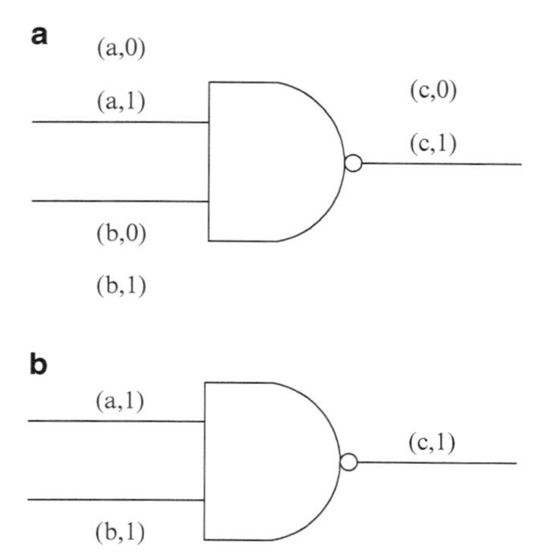

Fig. 2.7 Fault collapsing. (**a**) Uncollapsed fault set and (**b**) collapsed fault set

Usually, fault equivalence and fault dominance relationships are exploited only locally since the overhead for identifying all relationships is too high [Lio92]. The following example shows the local relationships of stuck-at faults at a NAND gate.

Example 2.3. Consider the 2-input NAND gate in Fig. 2.7a. At the output c and at each input a, b of the NAND gate, two stuck-at faults (s-a-0, s-a-1) are possible. Fault $(c,1)$ can only be detected when both inputs assume the non-controlling value 1. This is also necessary for detecting $(a,0)$ and $(b,0)$. Because this test pattern is the only existing test pattern for these faults, $(c,1)$, $(a,0)$ and $(b,0)$ are equivalent. Therefore, only one of them has to be targeted, e.g. $(c,1)$.

The fault set can be further reduced by using fault dominance relationships. A test for $(a,1)$ and $(b,1)$ necessitates the controlling value 0 at one input and therefore always detects $(c,0)$, but not vice versa. $(a,1)$ and $(b,1)$ dominate $(c,0)$. Consequently, $(c,0)$ needs not to be considered in the fault set. The collapsed fault set of the NAND gate is shown in Fig. 2.7b.

2.3.2 Delay

Due to the high operating speed of modern designs, delay testing is performed to detect timing defects as well as to guarantee that the design meets the performance specification. Clock signals are used to synchronize the inputs and all outputs are expected to assume their final value within a defined clock period. For this purpose, several delay fault models were developed to detect inconsistencies in the temporal behavior. Delay faults must be considered under different aspects than stuck-at faults, since the timing of the circuit is ignored during stuck-at testing. The most common delay fault models, i.e. the *Path Delay Fault Model* (PDFM) and the *Transition Fault Model* (TFM), are introduced in this section.

2.3.2.1 Path Delay

The PDFM [Smi85, LR87] is the most accurate delay fault model. A manufactured circuit is said to be free of timing defects if every path from a PI to a PO propagates its transitions in less time than the specified clock cycle. A *Path Delay Fault* (PDF) models a distributed delay on a structural path $\mathscr{P}$ from a PI to a PO of $\mathscr{C}$. A PDF occurs if the size of the delay defect on $\mathscr{P}$ exceeds the slack. The slack of a path is the difference between the clock cycle and the specified path delay. This fault model is suitable for detecting small as well as large delay defects.

Definition 2.4. A structural path $\mathscr{P}$ is defined as the sequence of gates $g_1, \ldots, g_k$, where g_1 is a PI or PPI and g_k is a PO or PPO. The path $\mathscr{P}$ must be complete. That means, each gate g_i on $\mathscr{P}$ with $0 < i < k$ must be an input of g_{i+1}. If a gate g_i is located on path $\mathscr{P}$, it is denoted by $g_i \in \mathscr{P}$.

A PDF occurs when the cumulative delay of all gates and signal lines along $\mathscr{P}$ exceeds the time for a specified clock cycle. Because the effects of a physical fault on the delay may be different for both types of transitions, two different fault types are modeled by the PDFM. According to the direction of the transition at the beginning of path $\mathscr{P}$, there is a rising PDF as well as a falling PDF for each structural path. A rising transition goes from logic 0 in the *initial* time frame to logic 1 in the *final* time frame. Analogously, a falling transition goes from logic 1 to logic 0. Formally, a PDF F is a tuple $(\mathscr{P}, t)$ where $\mathscr{P}$ is a structural path and $t \in \{\uparrow, \downarrow\}$ denotes the direction of the transition at g_1, i.e. $\uparrow$ is used for a rising transition and $\downarrow$ for a falling transition.

A test pattern for a PDF consists of two input vectors V_1, V_2 applied in two consecutive time frames t_1, t_2. The vector V_1 is applied in the initial time frame t_1 and places the initial transition value at the fault site. The vector V_2 is then applied at operating speed in the final time frame, launches the transition at g_1 and propagates it to g_k by sensitizing a path from g_1 to g_k. The transition has to arrive at g_k in the specified clock cycle before the sampling time. Otherwise, a delay fault is detected at $\mathscr{P}$.

However, if multiple delay faults are present in $\mathscr{C}$, a test pattern might not detect the fault because other delay faults may mask the targeted PDF. Therefore, sensitization criteria were developed to classify tests according to their fault detection abilities, i.e. the quality. A test is called *robust* if and only if it detects the fault independently of other delay faults in the circuit [Smi85, LR87]. *Non-robust* tests guarantee the detection of a fault if there are no other delay faults in the circuit [LR87]. Robust tests are more desirable to obtain since they provide a higher quality. Detailed discussions concerning the classification of PDF tests and further sensitization criteria for a more detailed differentiation, e.g. strong robust or functional sensitizable, can be found in [KC98].

Robust and non-robust tests differ in their constraints on the side inputs of $\mathscr{P}$. A side input s of $\mathscr{P}$ is an input of gate $g_i \in \mathscr{P}$ with $1 < i \leq k$ which is not on $\mathscr{P}$ ($s \notin \mathscr{P}$). The sensitization criteria for robust and non-robust test patterns are shown in Table 2.2. The values shown in this table correspond to the five-valued

Table 2.2 Sensitization criteria for robust and non-robust test patterns

Gate type	Robust Rising	Robust Falling	Non-robust
AND/NAND	$X1$	$S1$	$X1$
OR/NOR	$S0$	$X0$	$X0$

logic $\mathscr{L}_5 = \{S0, S1, X0, X1, XX\}$ originally proposed in [LR87]. The values $S0$ and $S1$ denote static values. The letter 'X' in a value's name denotes a don't care. This means that the initial value is don't care at $X0/X1$, while both, the initial and final value are don't care at XX.

All side inputs of $\mathscr{P}$ have to assume a non-controlling value in the final time frame for a non-robust test. The non-controlling value of a gate is the opposite value of the controlling value. The controlling value of a gate g is the logic value which, when assumed at any input of g, determines the output value of g regardless of the values on other signals, i.e. 0 for AND/NAND and 1 for OR/NOR.

The constraints on the side inputs for a robust test depend on the transition of the on-path input g_i. If the transition on g_i goes from the non-controlling value to the controlling value of gate g_{i+1}, the side inputs of g_{i+1} have to assume a static non-controlling value (denoted by $S0/S1$). A static non-controlling value at the side inputs guarantees that a delayed on-path transition of g_i cannot be masked by a transition or a glitch on the side inputs. If the on-path transition goes from the controlling value to the non-controlling value, the side inputs have to assume a non-controlling value only in the final time frame – as specified for the non-robust sensitization model. In this scenario, a delayed transition cannot be masked by the side inputs, because line g_{i+1} switches to the non-controlling value not until g_i and all side inputs switch to the non-controlling value. The following example demonstrates the different sensitization criteria.

Example 2.4. An example circuit is presented in Fig. 2.8. Let $(\mathscr{P}, \uparrow)$ with $\mathscr{P} = (i_2, a, b, d, o)$ be the PDF under test. The following constraints are needed for non-robust sensitization (see Fig. 2.8a). The side inputs of gate a and gate o are constrained to $X1$, while the side input of gate d is set to $X0$. This ensures that the non-controlling value is assumed in the final time frame. Additionally, a rising transition is assumed at i_2. This results in the following test pattern:

$$V_1 = \{i_1 = X, i_2 = 0, i_3 = X, i_4 = X\}$$

$$V_2 = \{i_1 = 0, i_2 = 1, i_3 = 1, i_4 = X\}$$

In contrast, the side inputs of gate d and gate o have to be sensitized by a static non-controlling value in the robust sensitization model (shown in Fig. 2.8b). This is due to the transition from the non-controlling value of the gate to the controlling value. Note that the transition is inverted after an inverting gate, e.g. gate d in this example. A robust test satisfying these constraints is for example:

$$V_1 = \{i_1 = 0, i_2 = 0, i_3 = X, i_4 = 1\}$$

$$V_2 = \{i_1 = 0, i_2 = 1, i_3 = 1, i_4 = 1\}$$

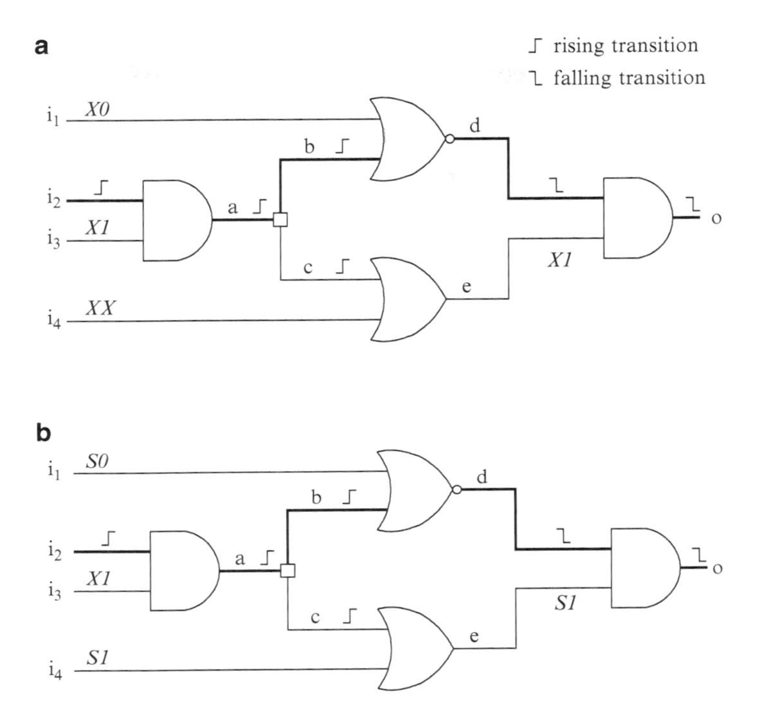

Fig. 2.8 Non-robust and robust sensitization. (**a**) Non-robust and (**b**) robust

The major limitation of the PDFM – which is the most accurate delay fault model – is the large number of paths in modern designs. The number of paths may be exponential in the number of gates. Performing ATPG for each single PDF is not practical due to the excessive computational effort and the resulting large size of the test set. Therefore, only a subset of all PDFs – faults on so-called critical paths – are considered for test generation in practice [LRS89, SP02].

2.3.2.2 Transition

In contrast to the PDFM, the TFM [BR83, LM86, SB87, WLRI87, Che93] is a local or "lumped" delay fault model. This fault model assumes a local delay defect, which affects only one single gate in the circuit. Similar to the PDFM, two different faults are associated with one fault site. The *slow-to-rise* fault models a delayed rising transition at the fault site, whereas the *slow-to-fall* fault models a delayed falling transition. The TFM is a specialization of the *Gate Delay Fault Model* (GDFM) [HRVD77, SB77, PR88] and has replaced the GDFM in industrial practice. The main difference between both fault models is that the GDFM includes the size of the delay defect while the TFM assumes that the delay defect is sufficiently large for detection.

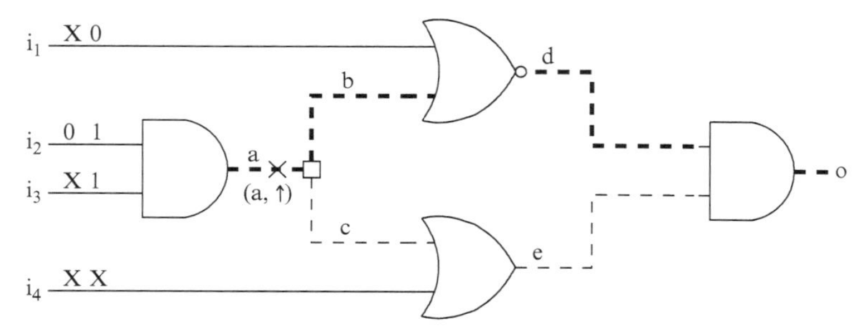

Fig. 2.9 Propagation paths – transition fault

Formally, a *Transition Fault* (TF) F is a tuple (f,t) with f as the affected gate or its outgoing connection, respectively, and $t \in \{\uparrow,\downarrow\}$. Since the size of the delay defect is not considered during test generation, the TFM is regarded as an idealized delay fault model. In order to test a TF at gate g, the fault must be activated at the affected gate and propagated to a PO. The activation path as well as the propagation path is chosen by the test generation algorithm. An activation path of a TF (f,t) is defined as a path from a PI to f which propagates the transition to the affected gate. A propagation path is the path which propagates the transition from the fault site f to a PO.

An advantage of the TFM is the similarity to the SAFM. Two consecutive time frames t_1,t_2 have to be considered for the TFM. The initial transition value is placed at the fault site in the initial time frame t_1. The transition is launched at operating speed and propagated to a PO in the final time frame t_2. This is done by logically sensitizing a path from the affected gate to a PO. In t_2, the TF behaves like the corresponding stuck-at fault, i.e. the slow-to-rise fault $(f,\uparrow)$ behaves like the s-a-0 fault $(f,0)$ and the slow-to-fall fault $(f,\downarrow)$ behaves like the s-a-1 fault $(f,1)$. The test vector V_2 is therefore identical to the pattern detecting the corresponding stuck-at fault. Figure 2.9 shows an example for the propagation of a TF.

Example 2.5. Consider the example circuit shown in Fig. 2.9, which has been already used to exemplarily present test generation for the PDFM. Assume that there is a TF $(a,\uparrow)$ in the circuit. The test generation algorithm can choose between two possible propagation paths:

$$\mathscr{P}_1 = (a,b,d,o)$$

$$\mathscr{P}_2 = (a,c,e,o)$$

A test for $(a,\uparrow)$ has to activate the transition at a and logically sensitize at least one of the shown propagation paths. The propagation along path $\mathscr{P}_2$ is not possible, because the final value on line d (caused by the transition on line a) would block path $\mathscr{P}_2$. A test for the TF $(a,\uparrow)$ using $\mathscr{P}_1$ as propagation path is for example:

$$V_1 = \{i_1 = X, i_2 = 0, i_3 = X, i_4 = X\}$$

$$V_2 = \{i_1 = 0, i_2 = 1, i_3 = 1, i_4 = X\}$$

Note that the test pattern is identical to the non-robust test pattern presented in Example 2.4. That is because the same activation path as well as the same propagation path is used.

Since the TFM is very similar to the SAFM, existing efficient solutions for the SAFM can be reused with slight modifications. However, the fault collapsing rules are more restrictive due to the consideration of two time frames [WLRI87]. As a result, the number of faults in the collapsed fault set is usually much higher for the TFM than for the SAFM.

The advantage of the TFM in contrast to other delay fault models is that the number of faults is linear in the number of gates and that, due to the similarity to the SAFM, existing ATPG solutions targeting stuck-at faults can be reused. The TFM is widely used in industrial practice providing a good fault coverage. As disadvantage, the TFM is not very accurate. This is because of the assumption that the increased delay is large enough to be detected by any propagation path which is not realistic. Furthermore, the general detectability of small delay defects, which becomes more and more important with the advancing technology, is low.

Several specializations of the TF were proposed. For example, the fault model ALAPTF (As Late As Possible Transition Fault) [GH04] was proposed to address the fault activation condition. It requires that the TF is launched via the longest robust segment ending at the fault site to accumulate small delay defects.

2.4 Classical ATPG Algorithms

Sophisticated ATPG algorithms were developed for the efficient generation of test patterns. An overview on the existing ATPG algorithms for each fault model described in the previous section is given below. Special attention is paid to the algorithms for stuck-at faults because the techniques introduced for this fault model are reused for other fault models as well.

2.4.1 ATPG for Stuck-at Faults

A symbolic test generation method for stuck-at faults based on the Boolean difference [SHB68] was presented in Sect. 2.3.1. However, the symbolic computation of test patterns turned out to be impractical for larger circuits. Subsequent ATPG approaches, see for example [SB91, SSAM93, Bec98], based on *decision diagrams* or *Binary Decision Diagrams* (BDDs) [Bry86], respectively, improve the run time behavior significantly but suffer from their excessive memory consumption when applied to modern circuits. Due to these shortcomings, these approaches did not achieve acceptance in industry. Therefore, this section concentrates on *path-oriented* ATPG algorithms which have been applied in industry for decades.

Table 2.3 Meaning of the values of $\mathcal{L}_5$

Value	Good circuit	Faulty circuit
0	0	0
1	1	1
D	1	0
$\overline{D}$	0	1
X	X	X

Algorithm 1 Outline of the D-algorithm

```
 1: Select_fault();
 2: while Untried_D-chain_exists() do
 3:    D-drive();
 4:    Consistency(); /* Involves decision making and backtracking if necessary */
 5:    if D-chain_is_justified() then
 6:       return TESTABLE;
 7:    else
 8:       Backtracking();
 9:    end if
10: end while
11: return UNTESTABLE;
```

2.4.1.1 D-Algorithm

The first seminal ATPG algorithm was the D-Algorithm [Rot66, RBS67]. The D-algorithm introduces an ATPG method based on the computation and propagation of D-values. The algorithm operates on a five-valued logic $\mathcal{L}_5 = \{0, 1, D, \overline{D}, X\}$ which is intended to represent the value of a signal line in both the correct or "good" circuit and the faulty circuit simultaneously. The concrete meaning of each value of $\mathcal{L}_5$ is presented in Table 2.3. The values D and $\overline{D}$ are used to represent signals which behave differently in the good and in the faulty circuit. The aim of the D-algorithm is to form a so-called *connected D-chain* (or short: *D-chain*) from the fault site to a PO. A D-chain is a complete path from the fault site to a PO, where all gates on this path assume either D or $\overline{D}$. By this, the propagation of the fault effect to an output is guaranteed. The outline of the D-algorithm is shown in Algorithm 1.

First, a target fault is selected and the initial D-value is injected at the fault site (line 1). That means D for a s-a-0 fault and $\overline{D}$ for a s-a-1 fault. Then, a D-chain is formed by selecting a propagation path from the fault site to a PO (line 3). This is done by driving a *D-frontier* towards the outputs. The D-frontier contains all gates which output values are X but one or more predecessors have a D-value, i.e. either D or $\overline{D}$. A gate g_d from the D-frontier is chosen to propagate the fault. The output of g_d is set to the corresponding D-value and the D-frontier is updated. This procedure is called *D-drive*.

After a D-chain was selected, the D-values on the D-chain have to be justified, i.e. a consistent input assignment has to be searched. This is done by the *Consistency* procedure in line 4. If such an input assignment is found, the algorithm returns with

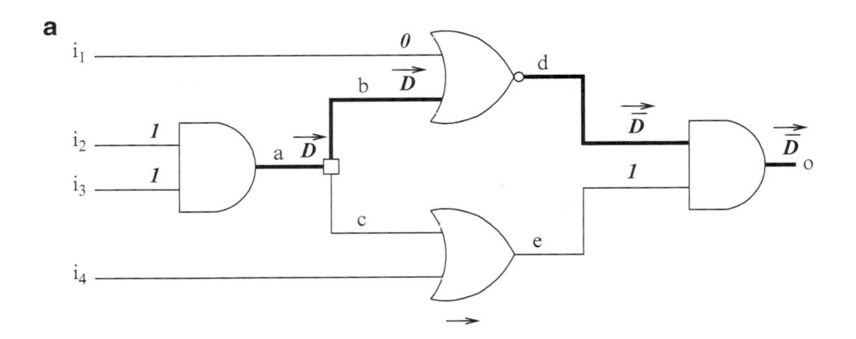

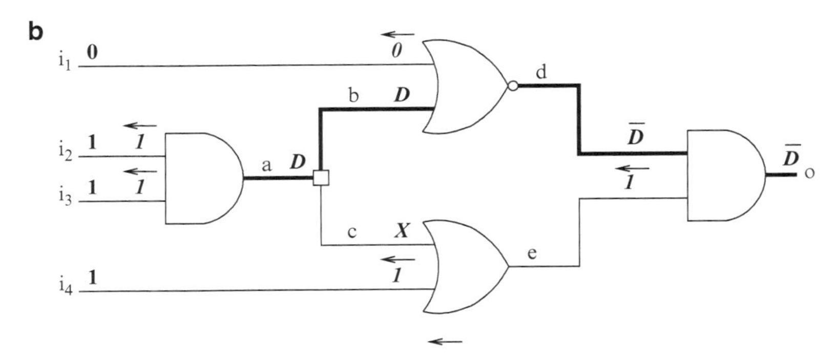

Fig. 2.10 D-algorithm steps. (**a**) D-drive and (**b**) consistency

the resulting test vector (line 6). Analogously to the D-frontier, the term *J-frontier* is introduced to keep track of the unjustified lines in the circuit [ABF90]. In contrast to the D-frontier, the J-frontier is driven towards the inputs. If the D-chain cannot be justified, *Backtracking* is performed and a new D-chain is computed (D-drive). If all potential D-chains were tried and no test pattern could be found, the fault is untestable.

Note that the Consistency procedure and the D-drive procedure involve decision making. At some point, implications are not possible and a decision has to be made. For example, to produce the value 1 at the output of an 2-input OR gate, either one input or both can assume the value 1. A decision tree is used to branch and bound through the search space. If a conflict situation happens, e.g. due to reconvergent paths, backtracking is performed to return to a previous decision point. The branch-and-bound procedure continues until either a test vector is found or the fault is proven to be redundant (untestable), i.e. the complete search space was traversed without finding a test. The following example demonstrates the procedure:

Example 2.6. Consider the circuit shown in Fig. 2.10. A s-a-0 fault on line *a* is targeted for test generation using the D-algorithm. Figure 2.10a shows the D-drive procedure. The D-frontier is driven from the fault site *a* to the output *o*. At the fanout gate, line b is selected. The resulting D-chain is (a,b,d,o). During the D-drive, value assignments necessary for the propagation of the D-values are gathered. The assignment $i_1 = 0$ is necessary for the propagation of the D-value from line b to line d, whereas the assignment $e = 1$ is necessary for the propagation from line d to the output o. The inputs i_2 and i_3 have to be assigned the value 1 to activate the fault.

When a D-chain is found, the Consistency procedure is invoked to justify the necessary assignments. This is shown in Fig. 2.10b. Here, an input assignment is searched which is consistent with the selected D-chain. In this example, only line e is not an input. Because line e is the output of an OR gate and the controlling value is assumed, it is sufficient that only one predecessor assumes the value 1. In this example, i_4 is chosen, whereas the line c is assigned to X. The generated test pattern is:

$$V_1 = \{i_1 = 0, i_2 = 1, i_3 = 1, i_4 = 1\}$$

The five-valued logic $\mathscr{L}_5$ introduced in [Rot66] was extended in [Mut76]. Here, four additional values were introduced (resulting in a nine-valued logic), which cover the cases where one (either the good or the faulty) value is known, but the other is unknown. By this, the nine-valued logic is more precise.

2.4.1.2 Improved Path-Oriented Algorithms

With the growing complexity of the circuits and the increasing number of re-convergences, the D-algorithm in its original form became very inefficient in the late 1970s. PODEM [Goe81] (*Path-Oriented DEcision Making*) was introduced to improve the performance of ATPG. The main reason of the inefficiency of the D-algorithm was the excessive number of backtracks which have to be performed. In PODEM, the test generation problem is formulated as a search of the n-dimensional 0–1 state space of primary input patterns of an n-input combinational circuit. Therefore, contrary to the D-algorithm, PODEM restricts the decision making to the PIs of the circuit. Thus, the complexity of test generation is reduced from $O(2^s)$ (D-algorithm) to $O(2^n)$ (PODEM), since decisions are carried out on a subset of signals only. This is possible, because all lines can be assigned by propagating the input assignment through the circuit. Here, s denotes the number of signals (including PIs, POs, and internal signals) in the circuit and n denotes the number of PIs – typically $n \ll s$.

Because decisions are restricted to the PIs, the *Consistency* procedure is replaced by a *Backtracing* procedure. Instead of directly assigning a value to a signal line (*objective*) and searching a consistent input assignment as done in the D-algorithm, the backtracing procedure traces a path from the objective signal line backwards to a PI. Backtracing is guided by heuristics and observability/controllability measurements (see e.g [Sne77, GT80]) and the selected PI is likely to help satisfying the objective. Then, forward implication is performed only. A further improvement of PODEM is the *X-path-check*. This is an early test whether the D-frontier still exists, i.e. there is at least one path of X's from the D-frontier to a PO. By this, conflicts can be detected earlier and the overall procedure is improved.

The algorithm FAN (FAN-out oriented test generation algorithm) [FS83] improves the efficiency of test generation by further reducing the number of backtracks and shortening the process time between single backtracks. Special attention is paid to fanout points respectively headlines. A headline is the output of a fanout-free sub-circuit. The following improvements are introduced in FAN.

- *Backward Implications* – In order to detect inconsistencies early, forward and backward implication is performed to assign as many values as possible.[4]
- *Unique Sensitization* – When the D-frontier consists only of a single gate, it often happens that all paths beginning at this gate go through one site or one specific path. If such a path is found, this path is partially sensitized in advance.
- *Multiple Backtracing* – Instead of backtracing a single path as suggested in PODEM, multiple paths are concurrently traced in a breadth-first manner. Decisions are carried out on headlines or fanout points of the circuit.

The FAN algorithm still forms the basis of many ATPG algorithms applied today in industry. However, several improvements were proposed to increase the efficiency of the test generation algorithm, e.g. [KM87, STS88, SA89, RC90, GB90, MGÖD90, WSGM90, KP93, MS94, HP99]. FAN is able to generate tests for a large number of "easy" faults very fast. Therefore, the majority of the proposed improvements aim to reduce the computational effort for "difficult" faults. In particular, significant improvements were achieved by the efficient determination of necessary assignments and the development of powerful learning concepts. The notion of *Learning* is used to describe the derivation of additional implications which cannot be obtained using the common forward and backward implication techniques alone. The basic learning concepts are outlined in the following.

2.4.1.3 Learning Concepts

The ATPG system SOCRATES (Structure-Oriented Cost-Reducing Automatic TESt pattern generation system) [STS88] improves the techniques introduced in FAN which are enumerated above. In particular, a global learning scheme was proposed. Indirect implications were identified in a pre-process carried out on the circuit structure using pre-defined rules. The additional "learned" implications denote relationships between signals which are not obvious. The application of these implications helps the algorithm to prevent "bad" decisions leading into a non-solution sub-space. As a result, the number of costly backtracks can be reduced and test generation is accelerated.

Example 2.7. An example (taken from [STS88]) is presented in Fig. 2.11. The figure shows a small circuit part. Simple forward implications are depicted in

---

[4]Backward implications were already an inherent feature of the D-algorithm, but discarded in PODEM.

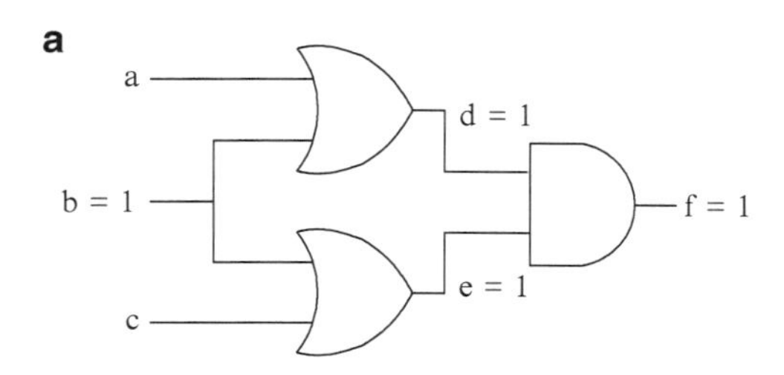

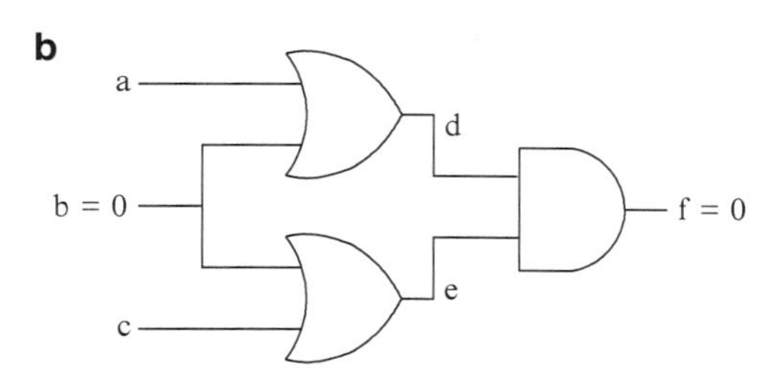

Fig. 2.11 Learning of indirect implications. (**a**) Direct implication of $b = 1$ and (**b**) indirect implication of $f = 0$

Fig. 2.11a. Assigning the value 1 to line b leads to an assignment $d = 1$ and $e = 1$, which in turn imply $f = 1$. Therefore, the direct implication $(b = 1) \rightarrow (f = 1)$ holds. The application of the law of contraposition yields the indirect implication $(f = 0) \rightarrow (b = 0)$, which is shown in Fig. 2.11b. This implication is indirect, because it cannot be obtained directly using backward implication.

An improved dynamic learning technique is presented in [SA89]. Here, the learning process is not performed once as a pre-process but dynamically during the search in each branching step, i.e. after each decision. As a result, the number of identified implications is much higher and test generation can be performed with fewer backtracks. As disadvantage, huge computational effort is needed for large circuits to perform dynamic learning in each branching step. The approach presented in [KP93] improves the dynamic learning technique by restricting the dynamic learning process to the "active area". This so-called *oriented dynamic learning* reduces the computational effort without sacrificing much important learned information.

An elaborated learning approach is shown in [KP94]. Here, the concept of *Recursive Learning* is presented. Instead of using a decision tree to keep track of the tried combinations of signal values, a recursive learning technique is used, i.e. recursively calling certain learning functions. The recursive learning technique is complete by itself. That means, all necessary assignments and logic relations between signals can be identified given enough recursions. As a result, a test can be found without any backtracks. However, recursive learning is very time-consuming when applied as stand-alone engine. In practice, this type of learning is combined with the FAN algorithm and only applied to leave a non-solution space as fast as possible.

The algorithm LEAP (LEvel-dependant Analysis in Path sensitization) presented in [MS94] introduces the techniques *failure-driven assertions* and *dependency-*

directed backtracking to prune search space in path sensitization problems such as ATPG. These techniques learn dynamically from conflicting signal assignments by analyzing the conflict and improving the backtracking procedure by leaping multiple decision points.

2.4.1.4 Boolean Satisfiability

In contrast to the ATPG approaches working on a structural circuit representation at gate level, methods based on *Boolean Satisfiability* (SAT) work on a Boolean formula. Different to algebraic methods (such as the Boolean difference method) which also work on a problem instance represented as a formula, SAT methods do not perform costly symbolic manipulations. SAT methods used for test generation can be grouped into two different categories which can be distinguished by the underlying data representation.

- *Conjunctive Normal Form* (CNF) – The CNF model represents the logic functionality of the circuit as a Boolean formula. Highly efficient SAT solvers were developed to evaluate such problems. A major reason for the robustness of modern CNF-based SAT solvers is the powerful conflict analysis (see Chap. 3 for more information). As disadvantage, structural information is lost, which is used by path-oriented ATPG algorithms to speed up the search.
- *Implication Graph* (IG) – An IG is a directed acyclic graph which nodes are the true or complemented signals or signal states, respectively. Implications are represented by edges between nodes. The advantage of the IG model over a topological model is that the IG abstracts from the pure structural description and combines structural and functional information in one model. However, IG-based methods do not incorporate conflict analysis techniques and do not benefit from the latest advances in SAT solving.

CNF-based SAT algorithms are e.g. TEGUS [SBS96], CGRASP/TG-GRASP [GSM99, MS97], PASSAT [DEF+08] and TIGUAN [CPL+10]. Because CNF-based algorithms are the main focus of this book, these approaches will be introduced more detailed in Chap. 4. A recent SAT-based ATPG approach is TIGUAN (Thread-parallel Integrated test pattern Generator Utilizing satisfiability ANalysis) [CPL+10]. State-of-the-art ATPG algorithms are single-threaded. The current trend in microprocessor design is the development of multi-core processors. Therefore, single-threaded ATPG algorithms do not use parts of the available computing power. However, TIGUAN uses the multi-threaded SAT solver MiraXT [LSB07] and can, therefore, utilize the performance of multi-core processors and control the number of threads. Since the use of multiple threads is a clear overhead for easy-to-test faults, TIGUAN proposes a two-stage procedure. In the first stage, TIGUAN is run only single-threaded within a short time interval. If no solution can be found within this short period, the approach utilizes thread parallelism for classifying the remaining hard-to-detect faults. In [CPE+09], TIGUAN is extended to be applicable in ATPG using dynamic compaction.

ATPG using an IG as problem representation was first introduced in TRAN [CAR93] and Nemesis [Lar92]. These first approaches modeled only binary relations between signals as implications. The IG model was incomplete, since ternary and k-nary ($k > 3$) relations had to be checked explicitly. This was improved by IGRAINE (Implication GRaph-bAsed engINE) [TGA00]. Here, the graph model was extended to ternary relations by using additional $\wedge$-nodes. k-nary relations can be transformed into multiple ternary relations. Graph analysis techniques are used to derive indirect implications. Furthermore, IGRAINE introduced a general method to transfer circuits represented in various logics to the IG model and provides a framework for efficient justification and propagation. IGRAINE was integrated into the ATPG tool TIP.

The IG model was further extended by SPIRIT (Satisfiability Problem Implementation for Redundancy Identification and Test generation) [GF02]. This approach improves the data structures in such a way that k-nary relations can directly be represented in the IG model. Furthermore, common structural ATPG techniques such as X-path check, unique sensitization and learning techniques are transferred to the IG model. Therefore, SPIRIT combines advantages of both worlds. However, the disadvantage of the proposed data structures is the large overhead for complex gates and gates with many inputs, respectively.

2.4.2 *ATPG for Delay Faults*

Classical approaches for ATPG for delay faults are presented in this section. Generally, many ATPG algorithms for delay faults are based on algorithms originally proposed for the SAFM. However, the quality aspect and the involvement of consecutive time frames require a different handling of the test generation procedures. First, test generation algorithms for the PDFM are presented. Then, algorithms for the TFM are reviewed.

2.4.2.1 Path Delay

Four different classes of ATPG algorithms for PDFs are identified:

- Algebraic algorithms
- Non-enumerative algorithms
- Structure-based algorithms
- SAT-based algorithms

Furthermore, the PDF test generation problem can be formulated as a stuck-at fault test generation problem as shown in [SBS92, GBA97]. However, this is not further discussed here, since the focus in this book is on direct PDF test generation formulation.

Algebraic algorithms do not work on the circuit structure, but on Boolean expressions, e.g. represented as BDDs. The approach in [BAA92] converts the circuit and the constraints that have to be satisfied for a delay test to BDDs. A pair of constraints is considered for each fault. Each constraint corresponds to one of the two time frames. Robust as well as non-robust tests are then obtained by evaluating the BDDs. The tool BiTeS [Dre94] constructs BDDs for the strong robust PDFM, i.e. for generating hazard-free tests. Instead of generating one single test pattern for a PDF, the complete set of tests for one fault is generated directly using BDDs.

Non-enumerative ATPG algorithms do not target any specific path, but generate tests for all PDFs in the circuit. Hence, the problem of the exponential number of paths in a circuit is avoided. The first non-enumerative ATPG algorithm was NEST [PRU95]. NEST considers all single lines in the circuit rather than the exponential number of paths and uses a greedy approach combined with fault simulation to generate many tests quickly. The approach does not perform well on poorly testable circuits due to the greedy nature. The approach ATPD [TK99] improves the greedy procedure of NEST in order to detect more PDFs by one test. RESIST [FPR94] does not use a greedy procedure but exploits the fact that PDFs are dependent, because many paths share sub-paths. Therefore, RESIST does not enumerate all possible paths, but sensitizes sub-paths between two fanouts, between input and fanout, or between fanout and output, respectively. The approach sensitizes each sub-path only once and, consequently, decreases the number of sensitization steps.

Both algebraic and non-enumerative approaches have serious shortcomings when applied to today's large circuits. As for the SAFM, algebraic algorithms suffer from their large memory consumption. Non-enumerative algorithms are outdated since a test set containing tests for all testable PDFs would be of excessive size. Furthermore, test generation is performed for critical paths only as described in Sect. 2.3.2.

Structure-based ATPG algorithms for PDFs work similar to the structural path-oriented algorithms for the SAFM. However, they differ in the following points. While for stuck-at fault test generation, path selection (propagation) and justification have to be performed, only justification has to be considered for PDF test generation since the path is completely specified. Additionally, algorithms for PDF test generation work on more complex multiple-valued logics. Those are needed because two time frames have to be considered. Furthermore, they were developed to ensure high quality delay tests. The justification procedures applied for path delay test generation are very similar to those applied for the SAFM. However, these procedures are performed on the basis of the dedicated multiple-valued logics. The following example shows such a multiple-valued logic.

Example 2.8. Figure 2.12 shows the Hasse diagram of the ten-valued logic proposed for robust test generation in [FWA93]. A value of this logic determines the signal behavior during two consecutive time frames. The value s denotes a static value, while $\overline{s}$ describes a non-static value. The don't care value is denoted by X and an unknown value is described by U. The lowest level shows the basic values (leaf nodes) of the logic and the upper levels present the composite values.

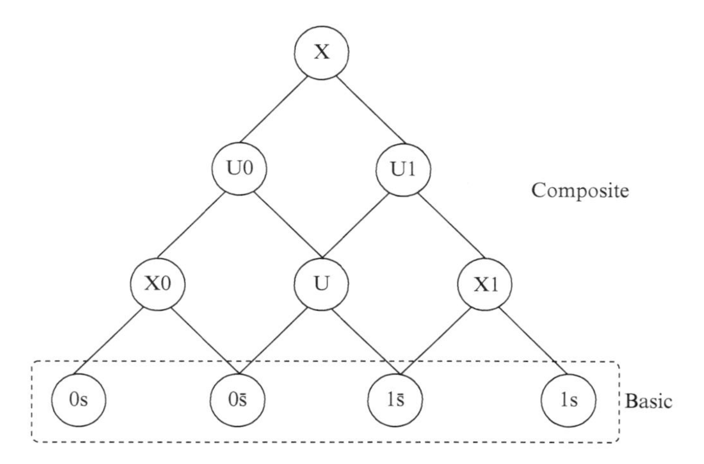

Fig. 2.12 Hasse diagram of ten-valued logic [FWA93]

Composite values are sets of basic values. For instance, the value $X0$ in the Hasse diagram describes the value set $\{0s, 0\overline{s}\}$ and the value X covers all other values contained in this logic.

The approach in [LR87] uses a five-valued logic to generate test patterns and introduces the general robust sensitization criterion (see Sect. 2.3.2). The underlying algorithm for the test generation is PODEM. The approach DYNAMITE [FFS91] is based on SOCRATES and proposes a ten-valued and a three-valued logic for robust and non-robust test generation, respectively. DYNAMITE provides a stepwise path sensitization procedure which is capable of proving large numbers of paths as untestable by a single test generation attempt. Consequently, DYNAMITE is very effective for circuits with a large number of untestable PDFs. The approach in [FWA93] enhances this scheme by using five different logic systems, e.g. a ten-valued and a 51-valued logic system, suitable for various test classes such as non-robust, robust and restricted robust. The work in [BAA98] is concerned with the derivation of an optimal set of logic states to minimize the number of backtracks during PDF test generation.

SAT-based algorithms work differently from those presented in this section since they do not work on the circuit structure. The problem of generating a test for a PDF is transformed to a Boolean SAT problem. The SAT problem is solved by a SAT solver. The SAT solution is then transformed into a solution of the original problem, i.e. a PDF test pattern. A detailed description of the basic SAT concepts is given in Chap. 3. The basic techniques for SAT-based ATPG are presented in detail in Chap. 4. The relevant SAT approaches are briefly described in the following for the sake of completeness.

The first SAT-based approach for PDF test generation was proposed in [CG96] where a seven-valued logic is used to generate robust tests for PDFs in combinational circuits. A Boolean encoding is applied for the transformation into a Boolean SAT problem. The underlying SAT engine was TEGUS. SAT-based learning techniques are applied in [KWMS00, CH05] to speed up PDF test generation. These approaches are all CNF-based algorithms. The IG-based approach IGRAINE [TGA00] was applied to target non-robust and robust test generation. The tool KF-ATPG is presented in [YCW04]. Unlike the above mentioned SAT-based approaches, KF-ATPG uses the circuit-based SAT solver presented in [LWCH03]. Therefore, KF-ATPG is able to exploit structural knowledge of the problem to speed up test generation. SAT-based approaches for PDF test generation will be described in more detail in Sect. 8.1.

2.4.2.2 Transition

Early approaches for the GDFM, e.g. [HRVD77, IRS88, PR88, SA89, MGÖD90, MC92, PM92, Mah93], suffer from the circumstance that the delay defect size involves costly calculations. Unlike for the GDFM, timing information does not have to be leveraged during the search for a TF test pattern. Concentrating on the logical behavior of the delay fault and disregarding the actual delays leads to more efficient test generation. Furthermore, existing test solutions can basically be reused with slight modifications since the TFM is very similar to the SAFM. Both "simplicity" and the reuse of existing algorithms is the reason for the wide use of the fault model – in spite of the lesser accuracy, i.e. the gross delay fault assumption.

The first deterministic ATPG algorithm for the TFM was presented in [LM86], where a modified D-algorithm was used for TF test generation in combinational circuits. The work presented in [Che93] deals with the particularities of testing TFs in sequential circuits. Since the existing stuck-at test generation algorithms are adopted for TF testing, TF testing consequently benefits strongly from the increasing efficiency of stuck-at test generators. However, as mentioned above, the efficiency comes at the expense of the accuracy and the ability to detect small delay defects. Stuck-at fault test algorithms are highly optimized to generate tests as fast as possible. Usually, shorter propagation paths are chosen since these paths are easier to sensitize.

This is contrary to the desired test quality requirement for the TFM where longer paths are preferred since small delay defects can be accumulated. This is called the *criticality* problem in [SPR02]. Therefore, many approaches were proposed to raise the quality level of TF tests. The gate delay test generator DTEST_GEN [PM92] can be considered as a first approach for test quality enhancement for the TFM. This approach can be seen as point of origin for several approaches with the aim to generate a TF test set with increased quality. In particular, a good trade-off between high quality tests, e.g. tests sensitizing longer paths, high fault coverage and efficient test generation, is desired.

The criticality problem of TF test generation is highly related to the problem of identifying critical paths for PDF testing. This issue is not included in depth in this book. The work [SPR02] presents the path-oriented TF test generator POTENT.

This approach uses structural PDF test techniques to search for the longest testable path passing through the fault site, as well as a PODEM-based test generation algorithm to generate a TF test pattern.

The approach TranGen [YCW04] employs a circuit-based SAT solver (see Sect. 3.5) to generate tests for path-oriented TFs. That means, the longest propagation path is identified and the ATPG algorithm will try to sensitize the path in a static manner, i.e. all side inputs have to assume static values. If the longest path is not sensitizable, the next longest path is chosen and so on. Conventional TF ATPG is performed if no path is sensitizable. Learning is done by identifying unsensitizable path segments which are stored in a circuit graph. However, the approach is based on path enumeration and suffers from the excessive number of paths in modern circuits.

An ATPG framework for generating high quality tests for TFs is presented in [KMT+06]. This framework is based on two methods: activation-first and propagation-first. Activation-first finds and fixes the longest functional activation path and then performs path propagation, whereas propagation-first finds and fixes the longest functional propagation path and performs fault activation afterwards. For each fault, it is dynamically decided which method is probably more promising. The underlying ATPG is based on SOCRATES.

Timing-aware ATPG is proposed in [LTW+06]. This approach works similar to the propagation-first method from [KMT+06]. Timing information is calculated in a pre-process. During test generation, the fault effect is propagated to a PO through the longest path first. Then in a second phase, all lines needed for justification of the propagation path are justified. Timing information is further used to maximize the transition arrival time at the fault site. As a result, the transition is launched as late as possible. The disadvantage of this method is the high run time as reported in [YCT08]. A SAT-based ATPG approach using timing information is proposed in [SCS+11]. Here, the path length calculation is encoded into the SAT instance and the enumeration of testable paths is focused. However, no transition-dependent delays are encoded using the approach.

2.5 Industrial Test Environment

Generally, a test pattern generation is not executed as a stand-alone program in industry, but is part of a larger test environment. Such a basic industrial test environment is exemplarily introduced in this section. The general flow in a test environment is given in Fig. 2.13. The flow can be divided roughly in the following three phases: netlist compilation, *Design-For-Test* (DFT) insertion and ATPG. The last one is described in more detail, because the techniques proposed in this book are applied in this phase.

First, the design is read in the *netlist compilation* phase from a hardware description language and compiled into a netlist consisting of a set of primitive gate types (see for example the gate types given in Fig. 2.3). Next, *DFT insertion* is performed. Test-specific components are inserted into the design in this phase.

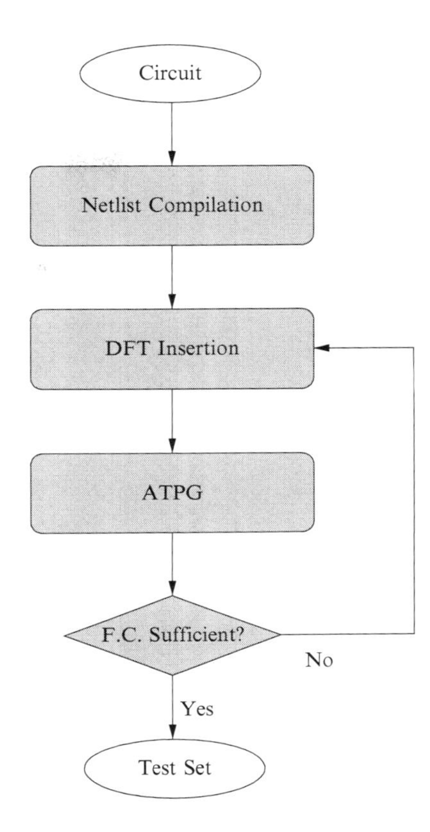

Fig. 2.13 Flow in industrial test environment

This includes for instance scan-chains and test points. After the DFT insertion, the ATPG phase is executed and a test set is generated. If the obtained fault coverage is too low, it is returned to the DFT insertion phase. Here, additional efforts can be undertaken to increase the fault coverage, e.g. test point insertion. Then, the ATPG phase has to be executed again.

In the ATPG phase, several methods interact to obtain a fast and effective ATPG system. Initially, a fault list is created containing all faults for a certain fault model which have to be tested. Random test patterns are generated and fault simulated (*Random Test Pattern Generation*, RTPG) . Fault simulation is very efficient – linear in the number of gates. All faults detected by these pattern can be removed from the fault list. By this, a large number of faults can easily be pruned. RTPG is typically performed until a defined fault coverage is reached. The remaining yet undetected faults are then targeted by *Deterministic Test Pattern Generation* (DTPG). Each generated test pattern is again fault simulated to find additional faults detected by this pattern. All detected faults are removed from the fault list until the fault list is empty.

Depending on the level of test set compaction which is executed, the interaction of the methods differs. Test set compaction is important, because a large test set

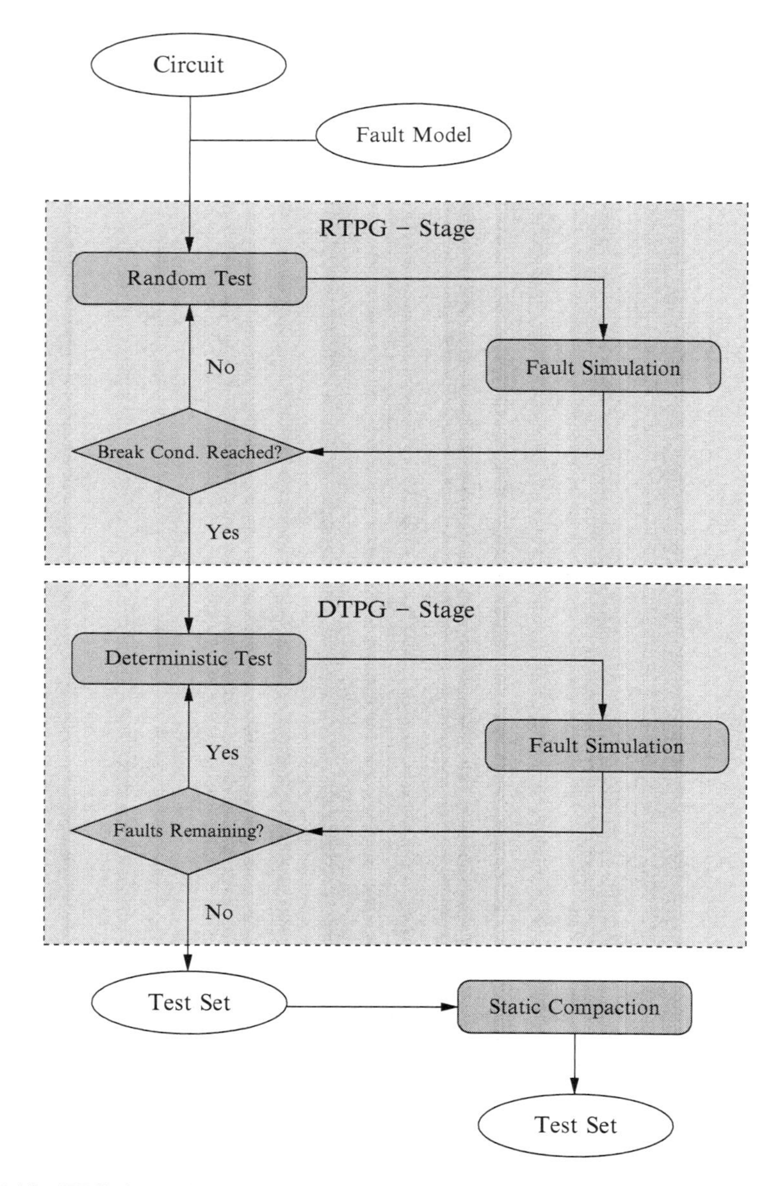

Fig. 2.14 ATPG phase – low compaction

leads to more test costs. In this description, it is distinguished between low and high compaction. Both flows are shown in the figures below. For low compaction (shown in Fig. 2.14), RTPG and DTPG are performed as described above. A static compaction method is applied after both stages have been finished. Here, those test patterns are removed, which only detect faults that are also detected by at least one

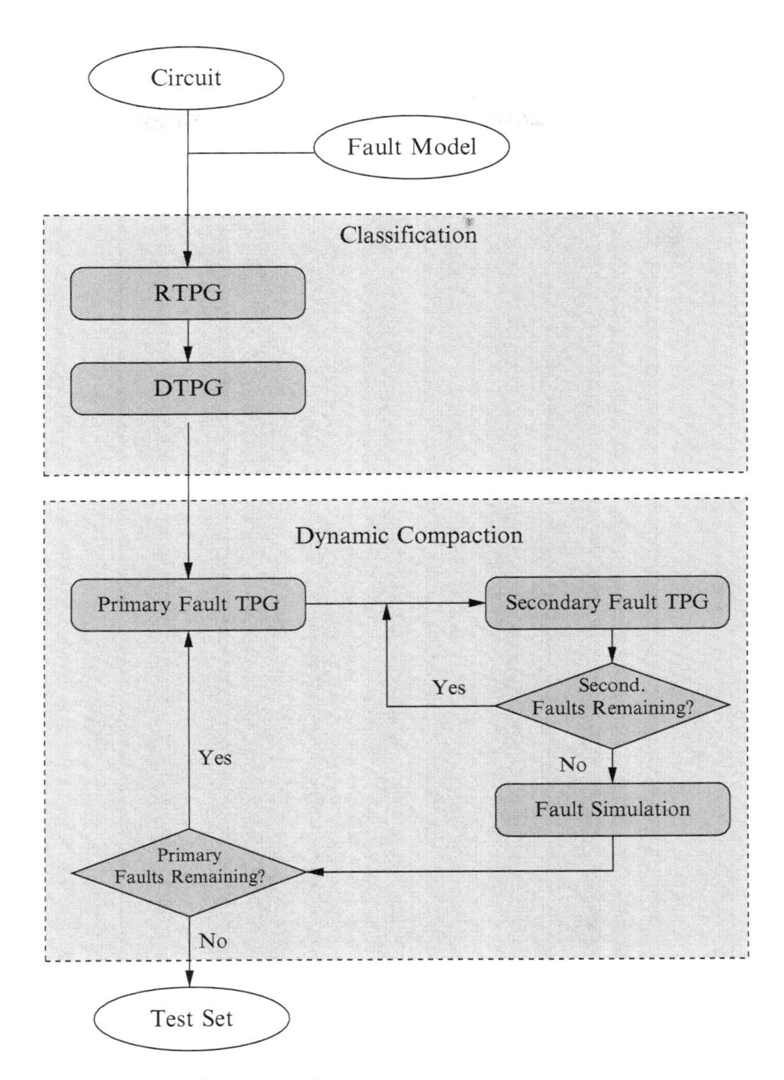

Fig. 2.15 ATPG phase – high compaction

other test pattern. Additionally, the tests are merged. Multiple test patterns which all have non-conflicting input assignments can be merged into a single test pattern. The number of don't care assignments should be generally as high as possible in order to achieve a good static compaction rate.

A dynamic compaction scheme is applied in the high compaction stage (shown in Fig. 2.15). The DTPG procedure is modified in order to generate a test pattern which detects one primary fault and as many secondary faults as possible. This is usually done by generating a test pattern for the primary fault, fixing the calculated input assignments and trying to detect other secondary faults under the assumption that the fixed inputs assignments hold. For detecting secondary faults, it is looped over all the yet undetected faults of the fault list. Again, a low number of don't cares is desirable to achieve a good compaction rate. In order to reduce the computational overhead of this method, all untestable and too-hard-to-detect faults are filtered out in a *classification stage*. In this phase, RTPG, DTPG and fault simulation interact as described above for static compaction. However, the test patterns are not retained.

Chapter 3
Boolean Satisfiability

This chapter introduces the *Boolean Satisfiability* (SAT) problem which is a central problem in computer science, since many real-world problems can be reduced to this particular problem. It further gives the basic information about SAT solving techniques which are necessary for the comprehension of this book and focuses on their application to circuit-oriented problems.

Section 3.1 introduces briefly the Boolean algebra and the basic notations. Section 3.2 gives the definition of the SAT problem and presents the classical DLL algorithm for solving SAT instances. Several improvements of this algorithm have been developed to increase the efficiency of SAT solvers. The most important improvements are described in Sect. 3.3. In this book, SAT algorithms are applied to circuit-oriented problems, i.e. the ATPG problem. Therefore, the basic knowledge about transforming a circuit-oriented problem into a SAT instance is presented in Sect. 3.4. Finally, a brief overview on circuit-oriented SAT solvers is given in Sect. 3.5.

3.1 Boolean Algebra

This section briefly describes the Boolean algebra and its basic notations and definitions to make the book self-contained. The *Boolean algebra* is the algebra of two values, i.e. Boolean values.

Definition 3.1. A *Boolean variable* x is a variable which can assume values from the set of Boolean values $\mathbb{B} = \{0, 1\}$. The Boolean value 0 is also denoted by FALSE and the Boolean value 1 can be written as TRUE.

Definition 3.2. A *Boolean function* f is a mapping of the form $\mathbb{B}^n \rightarrow \mathbb{B}^m$ where n, $m \in \mathbb{N}$. Usually, f is defined over a finite set of Boolean variables $X_n = \{x_1, \ldots, x_n\}$ and hence is also denoted by $f(x_1, \ldots, x_n)$.

S. Eggersglüß and R. Drechsler, *High Quality Test Pattern Generation and Boolean Satisfiability*, DOI 10.1007/978-1-4419-9976-4_3,

Table 3.1 Truth tables of logical operators (a) Unary (b) Binary

(a)

x	$\overline{x}$
0	1
1	0

(b)

x	y	$x \cdot y$	$x+y$	$x \oplus y$	$x \rightarrow y$	$x \leftrightarrow y$
0	0	0	0	0	1	1
0	1	0	1	1	1	0
1	0	0	1	1	0	0
1	1	1	1	0	1	1

Definition 3.3. A *Boolean expression* over X_n is an expression consisting of:

- Boolean variables
- Logical operators
  - Unary operator: $\bar{\cdot}$ (NOT)
  - Binary operators: $\cdot$ (AND), $+$ (OR), $\oplus$ (XOR), $\rightarrow$ (IMPLIES), $\leftrightarrow$ (EQUIVALENCE)
- Parentheses

The functionality of a logical operator can be described by a truth table. A truth table lists all possible values of the variables. The truth tables for the logical operators presented above are shown in Table 3.1. The evaluation of a Boolean expression follows an implicit ordering of the logical operators. Generally, if parentheses are used, expressions enclosed in parentheses are evaluated first. Otherwise, the ordering of the logical operators is as follows: NOT, AND, OR and XOR, IMPLIES and EQUIVALENCE.

In the Boolean algebra, the following properties hold:

$x+(y+z) = (x+y)+z$	$x \cdot (y \cdot z) = (x \cdot y) \cdot z$	associativity
$x+y = y+x$	$x \cdot y = y \cdot x$	commutativity
$x+(x \cdot y) = x$	$x \cdot (x+y) = x$	absorption
$x+(y \cdot z) = (x+y) \cdot (x+z)$	$x \cdot (y+z) = (x \cdot y)+(x \cdot y)$	distributivity
$x+\overline{x} = 1$	$x \cdot \overline{x} = 0$	complement

3.2 SAT Solver

The SAT problem was the first problem which was proven to be NP-complete [Coo71] and is defined as follows.

Definition 3.4. Given a Boolean formula $f(x_1,\ldots,x_n)$ with Boolean variables $x_1,\ldots,x_n$ consisting only of variables, parentheses and the logical operators AND, OR and NOT, the SAT problem is defined as the decision problem, whether there exists an assignment $a_1,\ldots,a_n$ for $x_1,\ldots,x_n$ such that $f(a) = 1$. If there exists one, a satisfying model is provided.

The task of a SAT solver is to find an assignment a such that $f(a) = 1$ or to proof that no such assignment exists. The Boolean formula f is called *satisfiable* (SAT) if such an assignment exists. In this case, the SAT solver provides the satisfying assignment. Otherwise, f is called *unsatisfiable* (UNSAT). SAT solvers typically work on a Boolean formula in *Conjunctive Normal Form* (CNF). A CNF is a conjunction (product) of clauses. A clause is a disjunction (sum) of literals. A literal is a Boolean variable in its positive or negative form.

In this book, a CNF or set of clauses is denoted by Φ. A clause is denoted by ω. Boolean variables and positive literals, respectively, are represented by lower Latin letters, e.g. a, whereas negative literals are represented by overlived lower Latin letters, e.g. $\overline{a}$. Indices are frequently used to distinguish the symbols when the same letters is used. A CNF Φ is satisfied if and only if each clause is satisfied. A clause ω is satisfied if and only if at least one literal contained in ω is satisfied. A positive literal l is satisfied if and only if the corresponding variable is assigned to 1, i.e. $l = 1$. A negative literal $\overline{l}$ is satisfied if and only if the corresponding variable is assigned to 0, i.e. $l = 0$. Note that an arbitrary literal with unknown polarity is typically denoted by λ.

The following example shows an example of a CNF.

Example 3.1.

$$\Phi = \underbrace{(a+b+\overline{c})}_{\omega_1} \cdot \underbrace{(\overline{a}+c)}_{\omega_2} \cdot \underbrace{(\overline{b}+c)}_{\omega_3}$$

The CNF Φ is a conjunction $(\cdot)$ of the clauses ω_1, ω_2 and ω_3. Each clause contains a disjunction $(+)$ of literals. At least one literal has to be satisfied in ω_1, ω_2 and ω_3 in order to satisfy Φ. A satisfying assignment is for example: $a = 0, b = 0$, $c = 0$.

3.2.1 DLL-Algorithm

An early algorithm for solving SAT problems was the *Davis-Putnam* (DP) algorithm [DP60]. This algorithm was based on the iterative application of resolution. The *Davis, Logemann and Loveland* (DLL) algorithm [DLL62] improves the procedure by replacing the resolution with a backtrack search and, by this, made it more space efficient.[1] The DLL algorithm still forms the basis of modern SAT solvers and is described in the following. Algorithm 2 shows the pseudo-code of the DLL algorithm as it is interpreted by modern SAT solvers, e.g. GRASP [MS99], Chaff [MMZ+01], BerkMin [GN02], MiniSat [ES04], and PicoSAT [Bie08b].

The DLL algorithm is a branch-and-bound search and works on a Boolean formula in CNF. The algorithm searches for an assignment which satisfies Φ by

[1]Due to the incremental development of the DLL algorithm and to acknowledge both works, the algorithm is often referred to as DPLL.

Algorithm 2 DLL-Algorithm

```
1: CNF Φ;
2: if Deduce() == CONFLICT then
3:    return UNSAT;
4: end if
5: while TRUE do
6:    if !Decide() then
7:       return SAT;
8:    else
9:       while Deduce() == CONFLICT do
10:         if Conflict_is_resolvable(); then
11:            Backtrack();
12:         else
13:            return UNSAT;
14:         end if
15:      end while
16:   end if
17: end while
```

iteratively choosing Boolean values, i.e. 0 or 1, for variables ("branch", *Decide()*, line 6). If all variables are assigned, i.e. no further decision can be performed, the algorithm determines SAT (line 7). Otherwise, one unassigned variable is picked heuristically and assigned with a different Boolean value. This variable assignment is then propagated (*Deduce()*, line 9). Here, all assignments which can be deduced from the decision and the current partial assignment are performed. This step is also called *Boolean Constraint Propagation* (BCP). Note that unsatisfiability can also be detected without making any decision. Therefore, before deciding the first variable, propagation is performed to detect trivial contradictions (line 2).

If a conflict is detected during BCP, the propagation is interrupted ("bound"). A conflict occurs if a deduced assignment is contradictory to an existent assignment or if all literals in a clause are false. If the conflict is resolvable, backtracking is performed (line 11) and the conflicting assignment is undone by inverting the last decision ("flipping"). Then, BCP is performed until a conflict is found again or no more implications can be derived. If the conflict cannot be resolved, the SAT instance is UNSAT (line 13).

It is important to point out that, due to reasons of efficiency, the condition to detect satisfiability in modern SAT solvers is the non-conflicting assignment of all variables and not – as in early approaches – the satisfaction of each clause.

3.3 Advanced SAT Techniques

The basic DLL algorithm was introduced in the previous section. However, this algorithm was substantially improved in the last decades. As a result, SAT solvers have become powerful reasoning engines and have been applied for many real world problems. The most important improvements to the original DLL algorithm are presented in this section.

3.3.1 Fast Boolean Constraint Propagation

The largest part of the solving time of a SAT solver is spent on BCP. Often more than 90% of the run time is used by the BCP process [MMZ+01]. BCP is carried out after each decision or variable assignment. The concrete task of BCP is to identify any additional variable assignment which can be deduced from the current (partial) assignment. The BCP process is mainly based on the iterative application of the *unit clause rule*. A *unit clause* ω_{unit} is a clause with one unassigned literal λ and apart from that only false literals or no other literals. The literal λ has to be true, because ω_{unit} has to be satisfied to satisfy the complete CNF. As a result, the assignment of λ can be implied.

Example 3.2. Consider the following clause ω:

$$\omega = (a + \overline{b} + \overline{c} + d)$$

Under the current assignment $b = 1, c = 1, d = 0$, ω is a unit clause, because all literals except a are false. Since a is unassigned, the assignment of a can be implied in order to satisfy ω: $a = 1$.

Watch lists are used for an efficient implementation of the BCP procedure. Such a watch list contains an entry for each variable x or literal λ_x used in the CNF. All clauses containing the literal λ_x are stored in the corresponding watch list entry. If x is assigned to a value, the BCP process can directly access these clauses and do not have to loop over all clauses to find additional assignments.

This procedure can be significantly improved by the *two-literal watch scheme* used in the SAT solver Chaff [MMZ+01]. This scheme was developed to reduce the number of propagation steps. Two arbitrary literals $\lambda_{i1}, \lambda_{i2}$ which are not false are watched in each clause ω_i. An implication in ω_i is only possible if either λ_{i1} or λ_{i2} evaluates to false and the other is unassigned. If another unassigned literal λ_{i3} exists, λ_{i3} is watched instead of the false literal. If no other unassigned literal exists, the other watched literal can be implied. A watch list is used which stores for each literal λ_i those clauses containing $\overline{\lambda_i}$. By this, only clauses with new false literals are considered during BCP. The procedure is still complete, because all implications are detected and conflicts can be detected likewise. However, SAT is possibly detected later since literals which evaluate clauses to true, are not watched. The following example demonstrates the procedure.

Example 3.3. Consider the following set of clauses:

$$\underbrace{(\overset{\downarrow}{a} + \overset{\downarrow}{\overline{b}} + c + d)}_{\omega_1} \cdot \underbrace{(\overset{\downarrow}{b} + \overset{\downarrow}{c} + d)}_{\omega_2} \cdot \underbrace{(\overset{\downarrow}{a} + \overset{\downarrow}{b} + c)}_{\omega_3}$$

Each initially watched literal is pinpointed by $\downarrow$. The corresponding watch list is given in Fig. 3.1a. Assume that the first decision is $d = 0$. There is no clause

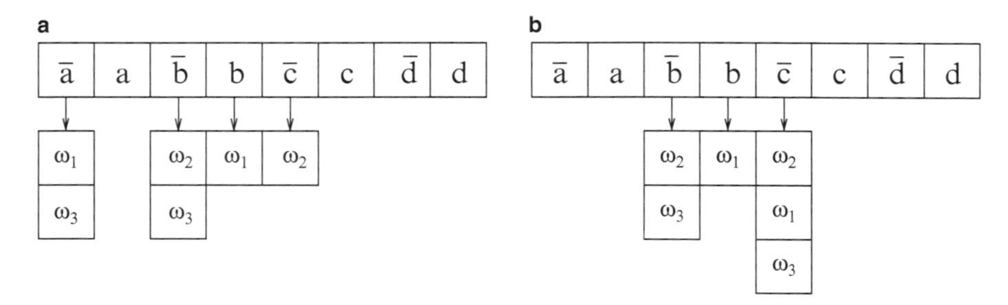

Fig. 3.1 Watch list example. (**a**) Initial and (**b**) Updated

associated to $\overline{d}$ in the watch list. Therefore, no implication can be deduced. Let $a = 0$ be the next decision. The watch list entry of $\overline{a}$ contains the clauses ω_1 and ω_3. Unassigned variables exist in both clauses. Therefore, the watched literals are changed and the watch list entries are updated. The changed watch list is shown in Fig. 3.1b. Now, literal c is watched in both clauses.

Assume that the final decision is $b = 0$. Then, ω_2 and ω_3 are considered for BCP. In ω_2, there is no other unassigned literal then c. Therefore, $c = 1$ can be implied. Because there is no conflict in ω_3 and the watch list entry of c does not contain any clauses, a satisfying assignment is found: $a = 0, b = 0, c = 1, d = 0$.

3.3.2 *Conflict Analysis*

One major reason for the robustness of modern SAT solvers is the ability to learn from conflicts. Conflict analysis was first proposed in the SAT solver GRASP [MS99]. Basically, the conflict analysis provides two benefits: *conflict clause* recording and *non-chronological backtracking*. First, conflict clause recording is described.

3.3.2.1 Conflict Clause Generation

If a conflict occurs during BCP, the conflict is analyzed and the reason for the conflict is identified. The reason of a conflict correspond to a subset of the current assignments which leads to the unsatisfied clause. A conflict clause is recorded and added to the clause database based on the identified reason. These conflict clauses are logically redundant and represent illegal partial assignments.[2] Considering

[2]The recording of conflict clauses corresponds to the derivation of implications in structural algorithms as described in Sect. 2.4.1. However, the concept is more powerful due to the underlying conflict analysis.

conflict clauses during BCP prevents the SAT solver at an early stage to enter a non-solution subspace. Without conflict clause recording, the SAT solver would enter this particular search space again under a different partial assignment.

A *directed implication graph* or simply *implication graph I* is maintained during the search in order to identify the reason of the conflict. Each vertex $v \in I$ denotes a variable assignment. Vertices without any predecessors represent decisions. Each implied assignment has the reasons of its assignment as predecessor. The edges are labeled with those clauses which caused the assignment. The reason of a conflict can be identified by a backward traversal of I. Generally, any cut through I separating the conflict from the decisions can be used as reason. When the reason is identified, a conflict clause ω_C can directly be created by excluding this partial assignment. Note that this procedure corresponds to the iterative application of resolution to the clauses involved into this conflict.

Generally, the size of ω_C is important for its deductive power. Therefore, heuristics are applied to find cuts with preferably few assignments. Here, the concept of *unique implication points* is important. More information about effective conflict clause generation can be found in [MS99, ZMMM01].

After the conflict is analyzed, the current decision level, i.e. one decision assignment and the resulting implication sequence, is erased from the search tree. As a consequence, ω_C becomes a unit clause and the unassigned variable can be implied. This is called *failure-driven assertion.* The use of failure-driven assertion replaces the "flipping" of the last decision variable. Note that the implied variable does not have to be the decision variable but can also be a variable which assignment was deduced. The procedure of conflict clause generation is demonstrated in the following example:

Example 3.4. Consider the CNF Φ shown in Fig. 3.2a. Assume that the decision heuristic chooses the following variable ordering during the search process: m = 1@1, $n = 1@2, e = 0@3, f = 0@4, h = 0@5, j = 0@6, g = 1@7$. The postfix $@x, x \in \mathbb{N}$ represents the decision level of the assignment. The resulting implication graph is shown in Fig. 3.2b. The vertices on the left hand side represent the decisions and the edges denote implications. Each edge has the clause in which the implication was detected as attribute. For instance, the implication $c = 1$ occurred in decision level 2 in clause ω_8. The implication is based on the assignments of m and n.

In decision level 7, the assignment of $g = 1$ causes the implications $d = 0, a = 0$ and $b = 1$. This results in a conflict in clause ω_1 because each literal evaluates to false under this assignment. The conflict is denoted by κ. A backward traversal of the implication graph is performed (starting at κ) in order to identify the reason for this conflict. An obvious reason can be obtained by the complete traversal to the decision nodes. Then, the resulting conflict clause would be: $\omega_C = (\overline{m} + \overline{n} + e + f + \overline{g})$. However, this conflict clause is rather long. A different cut through the implication graph can be used to derive a shorter conflict clause. Here, a more useful cut would go through c, d, e resulting in the conflict clause: $\omega_C = (\overline{c} + d + e)$. The method using graph cuts is based on the application of resolution. Let ω_1 be the

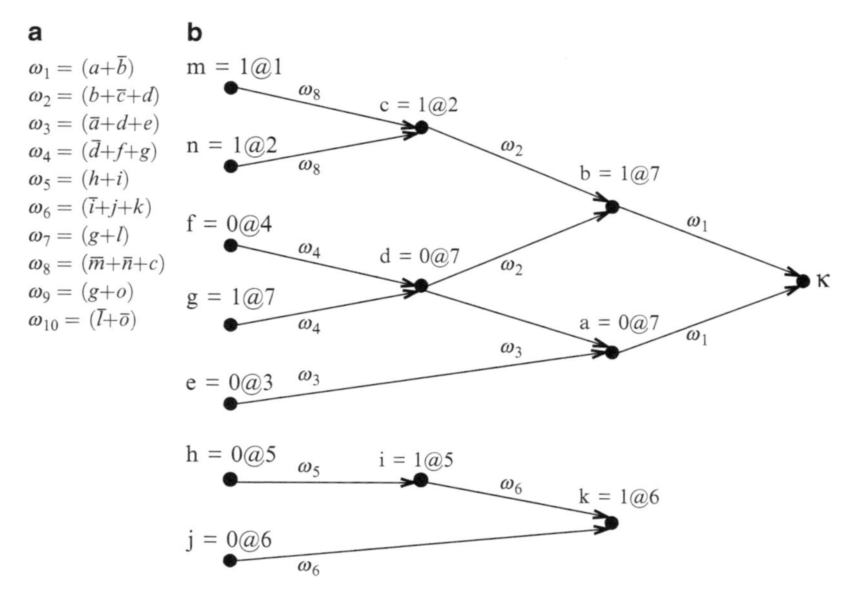

Fig. 3.2 Example CNF with implication graph. (**a**) CNF Φ and (**b**) Implication graph

unsatisfied clause and let $\odot$ be the resolution operator. The following resolution steps provide the conflict clause.

$$\underbrace{(a+\overline{b})}_{\omega_1} \odot \underbrace{(b+\overline{c}+d)}_{\omega_2} = \underbrace{(a+\overline{c}+d)}_{\omega_{r1}}$$

$$\underbrace{(a+\overline{c}+d)}_{\omega_{r1}} \odot \underbrace{(\overline{a}+d+e)}_{\omega_3} = \underbrace{(\overline{c}+d+e)}_{\omega_C}$$

By erasing the current decision level, i.e. undoing the assignments of a, b, d, g, $\omega_C = (\overline{c}+d+e)$ becomes a unit clause, since c and e are false literals. The assignment $d = 1$ is used as failure-driven assertion.

SAT solvers typically generate a large number of conflict clauses during their search. Therefore, inactive conflict clauses are deleted in order to avoid memory explosion. This is described in more detail in Sect. 3.3.3.

3.3.2.2 Non-chronological Backtracking

In the following, non-chronological backtracking (or conflict-directed backtracking) is described which is closely related to the use of conflict clauses. Backtracking has to be performed if a conflict is detected during the search process. If all possibilities

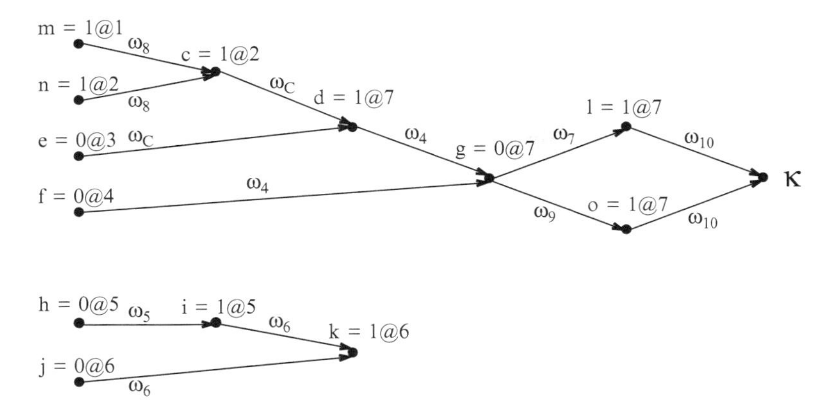

Fig. 3.3 Implication graph for non-chronological backtracking

of the current decision level $i > 0$ lead to a conflict, backtracking to the decision level $i-1$ is carried out. However, this so-called *chronological backtracking* does not involve the information derived from the conflict analysis. The conflict analysis provides information about the reason for the conflict. Therefore, the search process can directly backtrack from the current decision level i to the highest decision level j at which responsible assignments were made. All levels between i and j can be skipped. A demonstration of this procedure is given in the following example.

Example 3.5. Consider again the CNF Φ given in Fig. 3.2a and the conflict clause $\omega_C = (\overline{c} + d + e)$ from Example 3.4. After the conflict has been analyzed, the backtrack level is calculated. The backtrack level from Example 3.4 is level 7 and the failure-driven assertion is $d = 1$. However, this assignment leads again to a conflict. The resulting implication graph is shown in Fig. 3.3. The corresponding conflict clause, i.e. the identified reason for this conflict, is $\omega_{NCB} = (c + \overline{e} + \overline{f})$. The highest decision level of this reason is level 4. Therefore, backtracking is performed directly to level 4 and levels 5 and 6 are skipped. The assignments done in level 5 and 6 are detached from the actual conflict as can be seen in the graph.

3.3.3 *Conflict-Driven Heuristics*

The conflict analysis is crucial for conflict clause generation and non-chronological backtracking. Besides these techniques, the conflict analysis also has become a substantial part of the heuristics applied in a SAT solver, i.e. conflict clause deletion heuristic and decision heuristic.

3.3.3.1 Conflict Clause Deletion

A modern SAT solver generates a large number of conflict clauses during the search as described in Sect. 3.3.2. This leads to the problem of memory explosion especially for hard SAT instances. Furthermore, a huge clause database slows down the BCP procedure because all these clauses have to be processed. In order to avoid a significant run time increase due to the generation of an excessive number of conflict clauses, the conflict clause database is reduced periodically. Here, a heuristic is applied to delete only those clauses which are supposed to be unprofitable for the further search. For this purpose, the SAT solver BerkMin [GN02] introduced the *activity* metric for conflict clauses. A conflict clause is said to be useful for the further search if it is often involved in (recent) conflicts. Therefore, the activity of each conflict clause which is "touched" during the conflict analysis is increased. Those conflict clauses with a low activity are then deleted periodically. Moreover, the activity of a conflict clause is decreased in intervals ("aging"). By this, keeping conflict clauses during the complete search which were only useful at the beginning of the search is avoided.

3.3.3.2 Decision Making

Another important feature of the SAT solver is the decision heuristic. Because SAT is an NP-complete problem, the SAT solver could solve the problem in polynomial time if the solver always makes the "right" decision. Therefore, a "good" decision heuristic is necessary. The first decision heuristics were based on collecting information from the Boolean formula. For instance, the decision heuristic from the SAT solver GRASP is formulated as follows [MS99]:

"At each node in the decision tree evaluate the number of clauses directly satisfied by each assignment to each variable. Choose the variable and the assignment that directly satisfies the largest number of clauses."

The decision heuristic is improved by including information from the conflict analysis into the decision making. The first conflict-driven heuristic was VSIDS (*Variable State Independent Decaying Sum*) developed for the SAT solver Chaff [MMZ$^+$01] and improved in the SAT solver BerkMin [GN02]. This strategy is based on the observation that the search process is more robust if variables are picked which are often involved in conflicts. The aim of this decision heuristic is to satisfy conflict clauses in particular recently learned conflict clauses. For this purpose, each variable in each polarity, i.e. each literal, has a counter. The counter is incremented each time a newly generated conflict clause contains this literal. The literal with the highest counter is then picked as decision variable (with the corresponding polarity). The advantage of this procedure is that the heuristic is independent from the variable state and such has very low overhead in computing the decision variable.

The SAT solver BerkMin [GN02] improves this procedure by attaching an *activity* value to each variable. Not only those variables contained in conflict clauses are used for this heuristic but additionally all variables "touched" during the conflict analysis. The decision is then carried out on the (unassigned) variable with the highest activity value. The activity is periodically decreased ("aging") like in the heuristic for conflict clause deletion.

The impact of different decision heuristics was empirically evaluated on real-world problems in [Mar99]. A further result of these evaluations was that the decision result is indeed important but search space pruning techniques have a greater impact.

3.3.4 Incremental SAT

The *Incremental SAT* (ISAT) problem [Hoo93] is defined as follows:

Definition 3.5. Given a SAT instance Φ which is known to be satisfiable, the ISAT problem is defined as the question whether $\Phi \cdot \Phi_C$ for a given set of clauses Φ_C is also satisfiable.

In the first consideration [Hoo93], ISAT was restricted to solving one large problem Φ containing n clauses by dividing it into n partitions. The SAT instance Φ is then solved by incrementally adding one partition, i.e. one clause, at a time. By this, it is taken advantage of the previously computed solution and the complete solving process can often be accelerated.

The ISAT problem was later applied not only to one large SAT instance Φ, but to a series of similar SAT instances $\Phi_1, \ldots, \Phi_k$ [KWMS00, WKS01]. ISAT can be applied to $\Phi_1, \ldots, \Phi_k$ if all these SAT instances $\Phi_i = \Phi_p \cdot \Phi_s^i$ with $1 \leq i \leq k$ share a common prefix function Φ_p but have different suffix functions Φ_s^i. A SAT instance Φ_s is called extension of Φ_p [KWMS00].

The greatest benefit can be achieved if Φ_p is UNSAT. Then, all extensions of Φ_p can be determined as UNSAT without any computation. Otherwise, if a solution is found for Φ_p, the found solution can then be used for accelerating the search for a solution of the extensions. Here, the existing search tree of the solving process of Φ_p is used for searching a solution for the extension.

Another advantage of ISAT is the reuse of learned information [MS97, Sht01]. Let $\Phi_1 = \Phi_p \cdot \Phi_s^1$ and $\Phi_2 = \Phi_p \cdot \Phi_s^2$ be two SAT instances which share a common prefix Φ_p. Furthermore, let Φ_C^1 be the set of conflict clauses, which is deducible from Φ_1 (denoted by $\Phi_1 \vdash \Phi_C^1$) and let Φ_C^p be the set of conflict clauses deducible from Φ_p with $\Phi_C^p \subset \Phi_C^1$. The clauses contained in Φ_C^p were named *pervasive clauses* in [MS97]. More precisely, pervasive clauses are those conflict clauses which are deduced while solving Φ_1 but derived only from the subset Φ_p. The conflict clauses in Φ_C^p are also valid for Φ_2 and can therefore be reused for pruning search space in the solving process of Φ_2.

A problem of this technique is the isolation of Φ_C^p. Different techniques were introduced for this. For more information about the isolation of pervasive clauses, it is referred to [Sht01, ES03] and to Chap. 5 in this book.

3.3.5 Restarts

Restarts have been proposed to avoid spending too much run time in hard subspaces without solutions [GSK98, MMZ+01]. Often, SAT instances coming from industrial applications are easy to classify if a "good" decision ordering is known. However, it is hard to predict whether a decision ordering is "good" or "bad". When choosing a "bad" decision ordering, the solving process could spend too much time in hard subspaces. This is called *heavy-tailed cost distribution* [GSK98] or *heavy-tail behavior* [Bie08b]. In order to avoid this, a controlled amount of randomization is introduced in the solving process.

After a certain interval, the search tree is deleted and the current state of all variables is reset. Then, the search is restarted from the root of the search tree. The information learned so far is kept in the database [MMZ+01] which leads presumably to a different search path. Additionally, a certain (but small) amount of randomization is also introduced in the decision making. However, if restarts happen too often, the search process becomes incomplete, i.e. UNSAT cannot be proven anymore. The intervals between restarts are continuously increased in order to avoid such a situation. As a result, the SAT solver can finally exhaust the search space to proof unsatisfiability – provided that sufficient resources are available. The restart interval is usually determined by a certain number of conflicts which is increased after each restart.

Recently, a new adaptive restart strategy was proposed in [Bie08a] that measures the agility of the SAT solver based on the number of flipped assignments. Thus, the SAT solver decides dynamically when restarting the search is beneficial to avoid counterproductive restarts.

3.4 Circuit-to-CNF Transformation

SAT solvers are powerful reasoning engines and can be used as a black box for many types of problems. However, this requires the transformation of the original problem into a SAT instance. The transformation of a circuit into a Boolean formula in CNF, which is the input format of a SAT solver, is shown in this section. In particular, this is necessary for SAT-based ATPG.

The basic transformation of a circuit into CNF was presented in [Tse68, Lar92]. A combinational circuit $\mathscr{C} = (\mathscr{G}, \mathscr{S}, \mathscr{I}, \mathscr{O})$ consists of a set of gates $\mathscr{G}$, a set of connections $\mathscr{S}$, a set of inputs $\mathscr{I}$ and a set of outputs $\mathscr{O}$ as defined in Sect. 2.2. A Boolean variable x_s is assigned to each connection $s \in \mathscr{S}$ representing the value which is assumed on s for the transformation of $\mathscr{C}$ into CNF.[3] In case of a FANOUT gate, the outgoing branches can be denoted by the same variable.

[3]Note that throughout the book, the same name is often used for the connection and the Boolean variable.

Table 3.2 Truth table of NAND

a	b	c	$c = \overline{(a \cdot b)}$
0	**0**	**0**	**0**
0	0	1	1
0	**1**	**0**	**0**
0	1	1	1
1	**0**	**0**	**0**
1	0	1	1
1	1	0	1
1	**1**	**1**	**0**

Each gate $g \in \mathscr{G}$ is represented by a set of clauses Φ_g. These clauses ensure that the formula can only be satisfied by assignments which are consistent for the specific gate type. In other words, at least one clause contained in Φ_g evaluates to false if an assignment occurs which is not consistent with the function of g. The derivation of the CNF for a 2-input NAND gate is shown in the following.

Table 3.2 shows the truth table of a NAND gate g^{NAND} with inputs a, b and output c. The last column presents the result of the characteristic function of the gate denoting whether the respective assignment is consistent or inconsistent with the gate's function. Then, a clause is created from each inconsistent assignment by applying DeMorgan's theorem. The following clauses result:

$$\Phi_g^{\text{NAND}} = (a+b+c) \cdot (a+\overline{b}+c) \cdot (\overline{a}+b+c) \cdot (\overline{a}+\overline{b}+\overline{c})$$

This CNF representation is not minimal and allows for improvement which can be done by for example two-level logic minimization, e.g. ESPRESSO which is part of SIS [SSL+92]. A minimization of the clauses above results in the following final clauses:

$$\Phi_g^{\text{NAND}} = (a+c) \cdot (b+c) \cdot (\overline{a}+\overline{b}+\overline{c})$$

The CNF representation $\Phi_{\mathscr{C}}$ of the circuit $\mathscr{C}$ can be obtained by the conjunction of the clauses of all gates:

$$\Phi_{\mathscr{C}} = \prod_{i=1}^{n} \Phi_{g_i} \text{ with } n = |\mathscr{G}|$$

The circuit-to-CNF transformation is demonstrated by the following example.

Example 3.6. Consider the example circuit $\mathscr{C}$ depicted in Fig. 3.4. The CNF $\Phi_{\mathscr{C}}$ for the circuit $\mathscr{C}$ is as follows:

$$\Phi_{\mathscr{C}} = \underbrace{(a+g) \cdot (b+g) \cdot (\overline{a}+\overline{b}+\overline{g})}_{\Phi_g(\text{AND})}$$
$$\underbrace{\cdot\, (c+\overline{h}) \cdot (d+\overline{h}) \cdot (\overline{c}+\overline{d}+h)}_{\Phi_h(\text{NAND})}$$

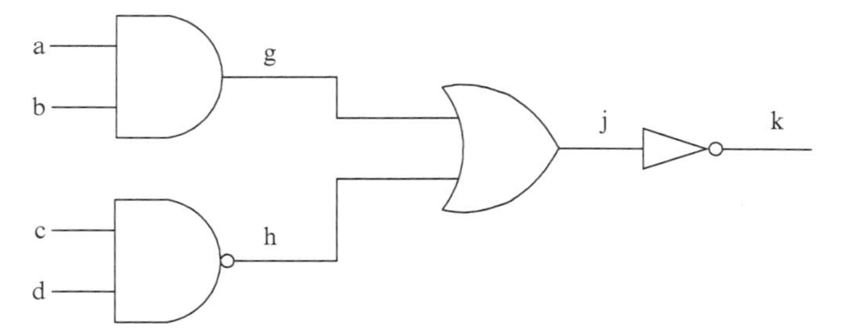

Fig. 3.4 Example circuit $\mathscr{C}$ for circuit-to-CNF transformation

$$\cdot \underbrace{(\overline{g}+j)\cdot(\overline{h}+j)\cdot(g+h+\overline{j})}_{\Phi_j(\text{OR})}$$

$$\cdot \underbrace{(\overline{j}+\overline{k})\cdot(j+k)}_{\Phi_k(\text{INV})}$$

The advantage of the gate-by-gate transformation is the linear size complexity. For a circuit with n gates (including inputs), n variables are needed and the number of clauses is in $O(3n)$ (assuming basic gates only). The disadvantage of this transformation is that the clauses are locally bounded and implications are only performed at a single gate. This effect is mitigated by use of conflict clauses in modern SAT solvers. However, several methods have been proposed to improve the CNF representation by a general pre-processing step, e.g. [Mar00, EB05]. Additionally, the circuit structure is exploited to improve the CNF representation e.g. in [Vel04, MV07].

3.5 Circuit-Oriented SAT and Observability Don't Cares

A disadvantage of CNF-based SAT solvers when applied to circuit-oriented problems such that the ATPG problem is the loss of structural information which happens during the circuit-to-CNF transformation. Structural information such as gate connectivity information or *Observability Don't Cares* (ODCs) are usually exploited by structural algorithms to accelerate the search or to reduce the search space, respectively. Therefore, it is desirable to make certain structural knowledge available to the SAT solver to speed up the search. CNF-based SAT solvers benefit significantly from the homogeneous structure of the Boolean formula which allows for the application of efficient data structures. The avoidance of overhead introduced by the insertion of structural knowledge is the key challenge. The benefit of structural knowledge has to be greater than the overhead. Different approaches were proposed which try to combine SAT techniques with structural information. These are briefly reviewed in the following.

The circuit-based SAT solver CSAT [LWCH03, LWC+03] does not work on a CNF representation but directly on circuit structure. Conflict clauses are represented as appended gates in CSAT. The implementation of CSAT is strongly influenced by the CNF-based SAT solver zChaff [MMZ+01], but the SAT techniques are transferred to the circuit representation. As an advantage, structural information can easily be exploited. CSAT strongly benefits from signal correlation guided learning. Signal correlations are identified during a simulation-based preprocessing and applied in two different ways: implicit and explicit.

In implicit learning, signal correlations are used to influence the decision strategy, whereas in explicit learning a correlated pair of signals is used to generate conflict clauses (recorded as gates). Such a pair of signals is assigned in a way that will most likely cause conflicts. The reasoning engine is then used to record conflict clauses (in the form of gates) derived from these assignments. The correlated pairs of signals are processed in topological order to reuse previously learned information. All learned information can then be utilized in the original problem instance. CSAT works especially well for problems with many signal correlations, e.g. equivalence checking problems.

QuteSAT [WLLH07] is similar to the circuit-based SAT solver CSAT but proposes a novel generic watching scheme on simple and complex gate types for BCP. Furthermore, it uses a new implicit implication graph to improve conflict-driven learning in a circuit-based SAT solver. QuteSAT focuses on efficient BCP techniques and data structures and does not contain structural techniques like implicit and explicit learning.

Hybrid SAT solvers work differently from circuit-based SAT solvers. Such a solver works partly on the circuit structure and partly on CNF. The approach presented in [GZA+02] processes only the original logic formula in the circuit representation. The authors argue that it is more efficient to treat a gate as a monolithic entry instead of a set of clauses since implications only require a single table look-up. Furthermore, circuit-based heuristics can be applied to speed up the search by retaining the circuit structure. In contrast, generated conflict clauses are processed in CNF representation, because they usually contain a large number of literals. Here, it is more efficient to use the two-literal watch scheme to avoid an excessive number of table look-ups.

The sequential SAT engine SATORI [IPC03] carries out BCP for the original clauses as well as for the conflict clauses completely on CNF. The circuit structure is used for efficient decision heuristics motivated by ATPG algorithms to guide the search. Structural information is used to reduce the assignments to primary inputs and state variables. Illegal states are recorded in the form of conflict clauses.

Since structural (ATPG) algorithms benefit significantly from the exploitation of ODCs, several SAT-based approaches were developed that are able to retain the information to detect ODCs and integrate them into the search process.

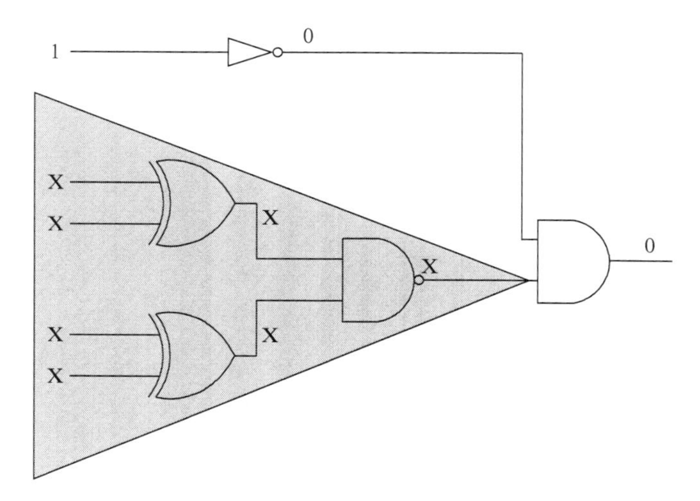

Fig. 3.5 Example circuit with ODCs

3.5.1 Exploitation of Observability Don't Cares

The notion of circuit *Observability Don' Cares* (ODCs) is as follows [FYM05]: In a circuit, some signal lines *S*, under certain conditions *Con* no longer affect the outputs of the circuit. The conditions *Con* are also called *ODC conditions*. Utilizing the existence of ODCs typically results in pruning portions of the search space because the assignment of *S* is don't care and has not to be justified.

Example 3.7. Consider the circuit shown in Fig. 3.5. Assume that the goal of the search is to produce the value 0 on the output. Then, it is sufficient to assign one input of the AND gate to 0. The other can be considered as don't care, because the controlling value is already assumed. As a result, the search process does not need to consider this part of the circuit since any assignment on this input cannot change the value of the output. The don't care part of the circuit is marked as a grey triangle. The ODC condition for this case is therefore that the other input of the AND gate assumes the controlling value 0.

The notion of ODCs in a SAT instance is similar [SVDL04]. Here, the assignment of a variable x, which does not influence the outcome of the current SAT solver goal, is considered as don't care. The solution space for x has not to be examined, i.e. branching on x is unnecessary. Since the ODC information is based on structural information of the circuit, common SAT solvers can not utilize this information per se.

Several approaches [GGYA01, Vel04, FYM05, SVD08] were developed which try to retain this information and utilize ODCs to speed up the search and prune parts of the search space. A hybrid SAT solver is presented in [GGYA01]. This approach retains the gate connectivity information and exploits them in the inner loop of the DLL algorithm to detect ODCs. In order to do this, a recursive backward traversal over all variables starting at the variables associated with the outputs is performed in intervals, e.g. in each decision level. Certain clauses are dynamically marked as inactive due to the ODC conditions during this traversal.

The approach presented in [SVD08] utilizes knowledge about the circuit structure to exploit ODCs. If the SAT solver assigns a value which is controlling for the successor gate g, all other inputs of g are marked as "lazy" which represents ODCs. As a result, decisions are not carried out on these variables, because an assignment would not influence the outcome of the search and the SAT solver does not enter this part of the search space.

The use of ODCs is improved in [FYM05], where it is integrated into the basic SAT techniques. Certain ODC literals – denoting ODC conditions – are added to clauses in the CNF during circuit-to-CNF transformation. This approach distinguishes itself from the methods presented above by applying the ODC information during the conflict analysis. The conflict analysis is altered in order to propagate the ODC conditions during the conflict clause generation. As a result, the generated conflict clauses also contain ODC information which can be applied during the remaining search process. The approach presented in [Vel04] does not change the search process of the SAT solver, but encodes the ODC information directly into the CNF. A variable is assigned to each logic block determining whether this logic block is unobservable or not. Constraints are added which propagate the effects of unobservability to other logic blocks in order to mark complete regions in the circuit dynamically as inactive.

Chapter 4
ATPG Based on Boolean Satisfiability

The first SAT-based ATPG approaches were proposed in the 1990s. However, these approaches did not become widely accepted because of some disadvantages such as the overhead for CNF transformation, missing support of multiple-valued logics and overspecified solutions. Additionally, existing structural ATPG algorithms were fast enough to cope with designs of that time.

However, Moore's Law is still valid and the size of the circuits roughly doubles every 18 months. Due to the ever increasing complexity of today's designs, structural ATPG algorithms reach their limits. Although being very fast for a large number of faults, the number of faults which can not be classified grows significantly. The high fault coverage demands of the industry cannot be satisfied more and more frequently. In contrast, the efficiency of state-of-the-art SAT solvers has been largely improved in the last decade as has been described in the previous chapter. As a result, there is a renewed interest in SAT-based algorithms and their application in the field of ATPG. Recently, the SAT-based ATPG approach PASSAT [SFD+05, DEF+08] was proposed. PASSAT benefits from the advances in SAT solving by integrating a state-of-the-art SAT solver as basic engine. Furthermore, PASSAT uses a multiple-valued logic and a Boolean encoding to cope with industrial particularities such as tri-state elements and unknown values.

This chapter starts with the description of the general SAT formulation for Boolean circuits as introduced in the 1990s as well as with the improvements developed since that time. Here, basic SAT formulations for the SAFM are presented. Section 4.2 deals with the particularities of industrial designs and motivates the use of a multiple-valued logic and a Boolean encoding as used by PASSAT. Section 4.3 presents the combination of a structural ATPG algorithm and a SAT-based ATPG algorithm in an industrial environment in order to exploit the advantages of both engines.

S. Eggersglüß and R. Drechsler, *High Quality Test Pattern Generation and Boolean Satisfiability*, DOI 10.1007/978-1-4419-9976-4_4,

4.1 SAT-Based ATPG for Boolean Circuits

The basic SAT formulation for an ATPG problem for faults in a Boolean circuit is given in this section. This formulation relies heavily on the circuit representation in CNF as described in Sect. 3.4. Section 4.1.1 presents the SAT formulation for the SAFM, while improvements on this SAT formulation are given in Sect. 4.1.2.

4.1.1 SAT Formulation: Stuck-at Fault Model

The SAT formulation for a single stuck-at fault consists of two parts. First, those circuit parts which are to be transformed in CNF have to be identified. And second, the fault detection has to be modeled in order to find a test pattern. Consider Fig. 4.1 for the description of the analysis for identifying the relevant circuit parts for one single fault. The starting point is the fault site. A depth-first search towards the outputs determines the *output cone* of the fault site. The output cone is the part of the circuit which could be influenced by the fault. This part is also called the *transitive fanout cone* of the fault site. Then, the *transitive fanin cone* of each output contained in the output cone is computed. The transitive fanin cone is also called *(structural) support*. Only those gates contained in this transitive fanin cone have to be considered when creating the SAT instance for generating a test pattern for this particular fault.

Next, the SAT formulation as introduced in [Lar92] is presented. Analogously to the construction of a Boolean difference (see Sect. 2.3), a fault free ("good") model of the circuit and a faulty model of the circuit are joined into a single circuit (see also Fig. 2.6) forming a miter-like structure. Duplicating the complete circuit or the transitive fanin cone, respectively, is a large overhead, because some parts of the

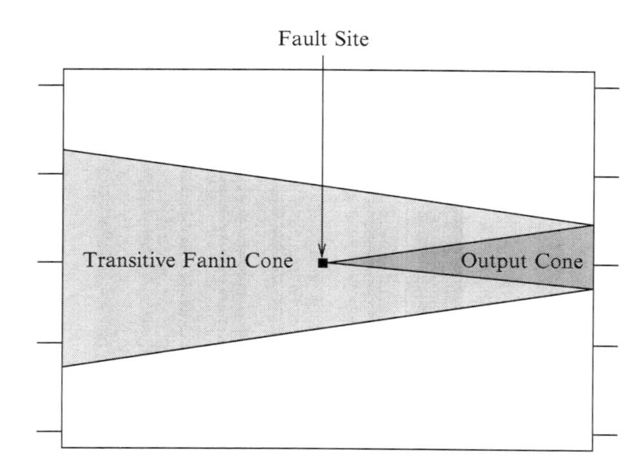

Fig. 4.1 Influenced circuit parts

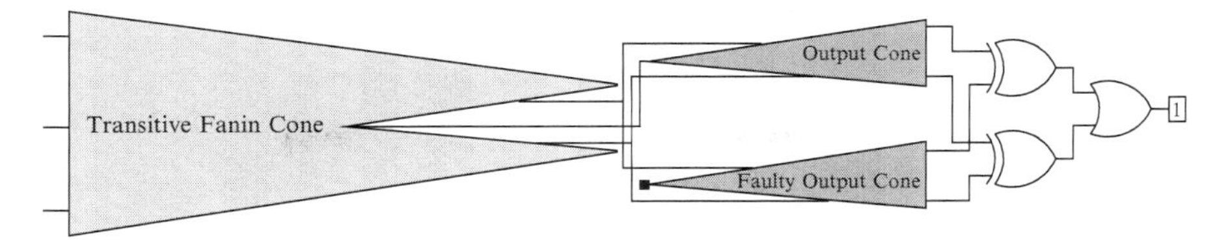

Fig. 4.2 SAT instance for ATPG

good and the faulty circuit are known to be identical. Therefore, only the output cone of the fault site has to be duplicated as it can be seen in Fig. 4.2. The remaining part can be shared between the good and faulty model.

This circuit is then transformed into a CNF by circuit-to-CNF transformation and the following constraints have to be added to the SAT instance to generate a test for the stuck-at fault (f, v):

- The fault site f in the faulty output cone has to be set to the faulty value v.
- The output of the miter, i.e. the output of the OR gate, has to be set to 1. This enforces at least one XOR gate to assume the value 1. That is, there is a difference between the output cone and the faulty output cone at least one output.

Forced by these constraints, every solution determined by the SAT solver corresponds to a valid test pattern for the fault (f, v). The fault is untestable if no solution can be found. The following example describes the SAT formulation for the SAFM by means of a concrete circuit and a concrete fault.

Example 4.1. Consider the example circuit shown in Fig. 4.3. Figure 4.3a shows the good circuit and Fig. 4.3b shows the same circuit containing a stuck-at fault $(a, 1)$. Since the value on the faulty line does not depend on $\mathscr{F}(a)$ anymore, a new line a' is created which is independent and assumes the faulty value. Since the faulty value can influence the output cone of a, the output o can also be affected (denoted by o'). The CNF Φ_G of the good circuit (depicted in Fig. 4.3a) is as follows:

$$\Phi_G = (\bar{i}_1 + \bar{i}_2 + a) \cdot (i_1 + \bar{a}) \cdot (i_2 + \bar{a}) \cdot (i_3 + i_4 + \bar{b}) \cdot (\bar{i}_3 + b) \cdot (\bar{i}_4 + b)$$
$$\cdot (a + b + \bar{o}) \cdot (a + \bar{o}) \cdot (b + \bar{o})$$

Since parts of the good and faulty circuit are identical, the faulty circuit does not have to be transformed entirely into CNF. Transforming the faulty output cone only is sufficient. This results in the CNF Φ_F. Since the faulty value is 1, the unit clause (a') is added to represent the behavior of the fault site.

$$\Phi_F = (a' + b + \bar{o}') \cdot (a' + \bar{o}') \cdot (b + \bar{o}') \cdot (a')$$

The CNF Φ_{BD} for the Boolean difference (shown as circuit representation in Fig. 4.3c) is a conjunction of Φ_G and Φ_F enhanced with an XOR gate to detect a

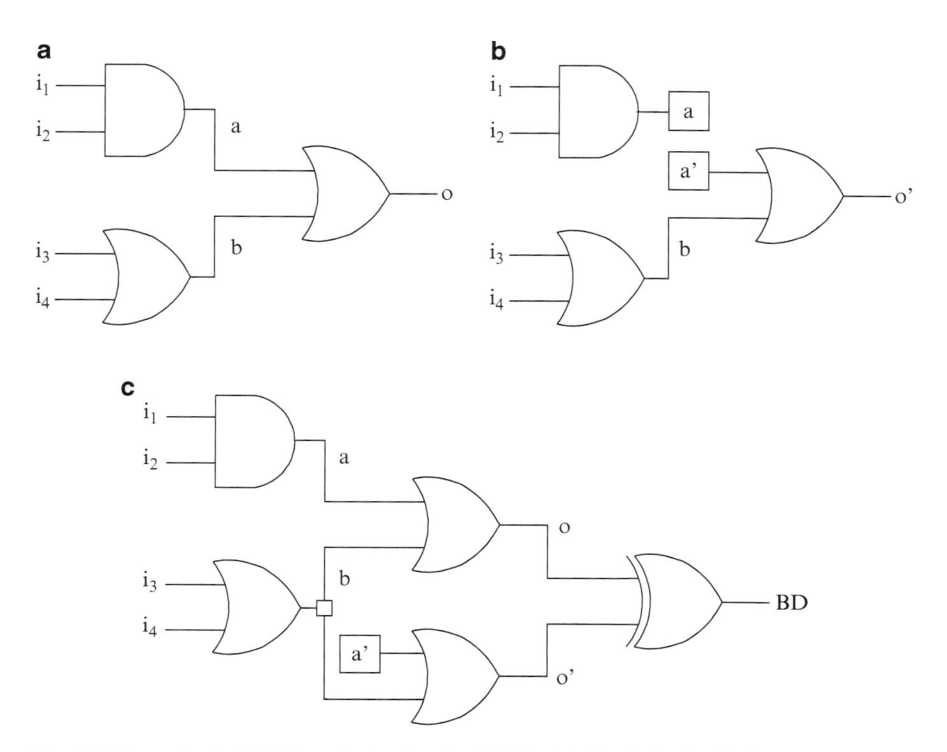

Fig. 4.3 Example for SAT formulation for the SAFM. (**a**) Good circuit, (**b**) faulty circuit, and (**c**) boolean difference

difference on output o. Since only one output is present in this example, there is no need for an OR gate. Finally, the unit clause (BD) sets the output of the XOR gate to 1. That is, the ATPG problem is to justify the value at the output. By this, all solutions where no difference can be observed at the output are excluded.

$$\begin{aligned}\Phi_{\mathrm{BD}} = &(\overline{i}_1 + \overline{i}_2 + a) \cdot (i_1 + \overline{a}) \cdot (i_2 + \overline{a}) \cdot (i_3 + i_4 + \overline{b}) \cdot (\overline{i}_3 + b) \cdot (\overline{i}_4 + b) \\ &\cdot (a + b + \overline{o}) \cdot (a + \overline{o}) \cdot (b + \overline{o}) \cdot (a' + b + \overline{o}') \cdot (a' + \overline{o}') \cdot (b + \overline{o}') \cdot (a') \\ &\cdot (\overline{o} + o' + \mathrm{BD}) \cdot (o + \overline{o}' + \mathrm{BD}) \cdot (o + o' + \overline{BD}) \cdot (\overline{o} + \overline{o}' + \overline{BD}) \cdot (\mathrm{BD})\end{aligned}$$

The SAT instance Φ_{BD} is given to a SAT solver. If the SAT solver determines SAT, a solution which corresponds to a test pattern is automatically yielded. Often, there is more than one satisfying assignment. SAT solvers typically return the first solution they are able to find. If the SAT solver determines UNSAT, the fault is untestable. In this example, the SAT solver provides the following satisfying assignment:

$$i_1 = 0, i_2 = 1, i_3 = 0, i_4 = 0, a = 0, a' = 1, b = 0, o = 0, o' = 1, \mathrm{BD} = 1$$

The test pattern T derives from the assignment of the input variables:

$$T = \{i_1 = 0, i_2 = 1, i_3 = 0, i_4 = 0\}$$

4.1.2 SAT-Based ATPG Techniques

The procedure to create a SAT instance for ATPG presented above is very simple and allows for improvements. Major improvements which speed up the search and enhance the robustness are presented in this section. Since the improvements are associated with specific SAT-based ATPG approaches, this section is structured in chronological order with respect to these approaches.

4.1.2.1 TEGUS

TEGUS [SBS96] improves the SAT formulation by encoding structural information into the SAT instance. A drawback of the SAT formulation using Boolean difference is that the good and the faulty part of the circuit (or the output cone, respectively) are distinct and only connected at the cut to the shared part and the outputs. This leads to computational overhead and an increased number of backtracks, since the observation of the fault can only be determined at the outputs. The concept of D-chains proposed in the original D-algorithm [Rot66] (see Sect. 2.4.1) is encoded into the SAT instance in TEGUS. A D-chain is a path from the fault site to an output on which the fault can be observed.

An additional variable (*d-variable*) is introduced in order to encode the concept of D-chains into the SAT instance. Therefore, the following three variables are used for each gate g in the output cone of the fault site[1]:

- g^c denotes the value in the good (correct) circuit.
- g^f denotes the value in the faulty circuit.
- g^d denotes, whether gate g is on a D-chain.

Additional implications and constraints have to be included in the SAT instance to compute the value of g^d.

- If g is on a D-chain, there must be a difference in the good and faulty circuit at g: $g^d \rightarrow (g^c \neq g^f)$. In terms of clauses, this implication is formulated as

$$(\overline{g}^d + g^c + g^f) \cdot (\overline{g}^d + \overline{g}^c + \overline{g}^f)$$

[1]Note that the use of these three variables makes the duplication of the output cone dispensable since it is handled implicitly.

- If g is on a D-chain, at least one successor of g must be on the D-chain as well. Let $h_1, \ldots, h_n$ be the successors of g, then $g^d \rightarrow \sum_{i=1}^{n} h_i^d$ must hold. This is formulated in CNF as follows:

$$(\overline{g}^d + h_1^d + \ldots + h_n^d)$$

- Given the faulty gate f, the unit clause (f^d) has to be included in order to force the calculation of a D-chain.

Adding these additional clauses to the SAT instance seems to be an overhead on first consideration. However, the advantage of this D-chain encoding is that the corresponding gates in the good and in the faulty parts are directly linked. This leads to possible implications between both parts in order to favor fault propagation. Furthermore, the propagation of the fault effect along one gate triggers the propagation along one of the successors until a PO is reached. As a result, costly backtracks are avoided, because the masking of the fault effect is detected much earlier. Note that the encoding of the D-chain makes the XOR construction of the Boolean difference method, i.e. the logic to compare the output's values, redundant.

Example 4.2. Consider again the good circuit given in Fig. 4.3a and the faulty circuit presented in Fig. 4.3b. The CNF formula Φ_G represents the structure of the good circuit and the faulty part is represented in CNF by Φ_F:

$$\begin{aligned}\Phi_G &= (\overline{i}_1^c + \overline{i}_2^c + a^c) \cdot (i_1^c + \overline{a}^c) \cdot (i_2^c + \overline{a}^c) \cdot (i_3^c + i_4^c + \overline{b}^c) \cdot (\overline{i}_3^c + b^c) \cdot (\overline{i}_4^c + b^c) \\ &\quad \cdot (a^c + b^c + \overline{o}^c) \cdot (a^c + \overline{o}^c) \cdot (b^c + \overline{o}^c) \\ \Phi_F &= (a^f + b^c + \overline{o}^f) \cdot (a^f + \overline{o}^f) \cdot (b^c + \overline{o}^f)\end{aligned}$$

The structural information, i.e. the encoding of D-chains, is then given by Φ_D:

$$\Phi_D = \underbrace{(\overline{a}^d + a^f + a^c) \cdot (\overline{o}^d + o^f + o^c) \cdot}_{\text{embedding D-chains}} \underbrace{(\overline{a}^d + o^d)}_{\text{fault propagation}} \cdot \underbrace{(a^f) \cdot (a^d)}_{\text{fault modeling}}$$

Finally, the test for the stuck-at fault $(a, 1)$ can be derived by evaluating the formula:

$$\Phi_{(a,1)} = \Phi_G \cdot \Phi_F \cdot \Phi_D$$

In addition to embedding structural information, TEGUS statically computes global implications in a pre-process. This is similar to the procedure introduced by the structural ATPG system SOCRATES [STS88]. However, the computation of global implications proposed in TEGUS is more powerful, i.e. more implications are found, using the SAT-based branch-and-bound procedure. As a result, the search space is reduced.

4.1.2.2 CGRASP

In contrast to TEGUS, the SAT-based ATPG approach CGRASP [GSM99] or TG-GRASP [MS97], respectively, introduces different layers. In addition to the layer of the SAT algorithm, i.e. the SAT solver, a new circuit layer is proposed that maintains circuit-related information. The advantage of using layers is that the underlying SAT solver can easily be exchanged if a more efficient one is available. The SAT solver GRASP [MS99] is used in CGRASP. GRASP incorporates conflict analysis and the use of conflict clauses. Therefore, CGRASP does not explicitly compute global implications as TEGUS does. CGRASP dynamically learns implications or conflict clauses, respectively, during the search process.

The circuit layer is intended to resolve the following problems: inaccessibility of structural information and over-specified test patterns. The following information is contained in the circuit layer for each gate g in the circuit:

- All predecessors of gate g.
- All successors of gate g.
- A threshold value $v_0(v_1)$ denoting the number of assignments necessary to justify $0(1)$ on gate g.
- A counter $c_0(c_1)$ denoting the number of actual assignments of the inputs of g in order to determine whether a gate is justified.

The task sharing of the SAT layer and the circuit layer is as follows: Value consistency and conflict analysis are handled by the SAT layer, while justification is handled by the circuit layer. Using the structural information, the circuit layer is able to maintain a J-frontier [ABF90] known from structural ATPG algorithms. The J-frontier is used for determining *syntactic satisfiability* [MS97].

By definition, a SAT instance is satisfied when all clauses are satisfied. However, this leads to over-specified test patterns which are disadvantageous for test compaction as explained in Sect. 2.5. The use of a J-frontier allows for stopping the search when the target fault and the propagation path is justified although not all clauses are yet satisfied. This modified break condition is known as syntactic satisfiability.

Another important concept introduced in CGRASP is the use of *pervasive clauses* which is closely related to global implications. During the search for a solution, the SAT solver generates conflict clauses. If a generated conflict clause ω_C is independent of the current target fault, i.e. ω_C only depends on the function and structure of the circuit, ω_C can be reused to prune search space for all other target faults as well. More information about pervasive clauses can be found in Chap. 6 in this book.

The significant overhead for mapping the ATPG problem into a CNF is identified in [MS97, GSM99] as a major problem for SAT-based ATPG where future research is needed.

4.2 SAT-Based ATPG for Industrial Circuits

All approaches presented so far can be used for Boolean circuits only, i.e. all signals are able to assume the Boolean values 0 or 1. However, considering Boolean values only is insufficient for ATPG in an industrial test environment. Industrial circuits typically contain elements with non-Boolean behavior. Therefore, ATPG tools have to be able to cope with these kind of gates as well. Structural ATPG approaches usually apply implication procedures based on multiple-valued logics (see [vdL96] for more information).

This section gives a brief overview how multiple-valued logics can be handled by SAT-based ATPG algorithms which work on a Boolean level. The SAT-based ATPG approach PASSAT [DEF$^+$08] uses a multiple-valued logic and a Boolean encoding to represent additional states. Section 4.2.1 shows the multiple-valued logic and gives information about the Boolean encoding used. More information can be found in [DEFT09].

4.2.1 *Multiple-Valued Logic*

In order to consider particular properties of industrial circuits, additional values have to be used to model the correct logical behavior, i.e. the values U and Z.

The value U denotes an unknown value. Since the assignment of some inputs of an industrial circuit may not be directly controllable during the test, e.g. because they are connected to a memory block, these inputs have to be assigned to the unknown value U. Note that the value U is not a don't care value but assumes a fixed state. Unknown values have to be considered in a special way since they are able to influence the fault propagation and justification.

The value Z denotes the electrical state of high impedance. The state of high impedance is important, e.g. for bus structures where multiple sources drive a single signal. Gates which can handle the state of high impedance are called tri-state elements in the following.

Considering unknown values and the state of high impedance, the following four-valued logic is derived:

$$\mathscr{L}_4 = \{0, 1, U, Z\}$$

Table 4.1 shows the truth table of an AND gate represented in $\mathscr{L}_4$. Other gate types are defined analogously. A Boolean encoding has to be used in order to apply a Boolean SAT solver to a circuit-oriented problem where the circuit is represented in a multiple-valued logic. In the following, the usage of Boolean encodings is introduced.

A Boolean encoding η is used to transform a multiple-valued problem in a Boolean problem. The value of each signal is represented by one Boolean variable in a purely Boolean circuit. In a multiple-valued circuit representation, one Boolean

Table 4.1 Truth table for an AND gate in $\mathscr{L}_4$

AND	0	1	U	Z
0	0	0	0	0
1	0	1	U	U
U	0	U	U	U
Z	0	U	U	U

Table 4.2 Boolean encoding for $\mathscr{L}_4$

$\eta_{\mathscr{L}_4}$	0	1	U	Z
c	0	1	1	0
c^*	0	0	1	1

Table 4.3 CNF representation of an AND gate using $\eta_{\mathscr{L}_4}$

$$(\overline{a}+\overline{b}+o)\cdot\ (a^*+b^*+\overline{o}^*)\cdot(\overline{a}^*+\overline{b}^*+o^*)\cdot \\ (a+a^*+\overline{o})\cdot(b+b^*+\overline{o})\cdot\ \ (\overline{a}^*+b+o^*)\cdot \\ (\overline{a}+\overline{b}^*+o)\cdot(o+\overline{o}^*)$$

variable is insufficient. More Boolean variables have to be used to represent all values. A logarithmic encoding was chosen for the four-valued logic $\mathscr{L}_4$ and two Boolean variables c, c^* are used to encode all four values and represent the signal's value.

The encoding is not unique. Using a logarithmic encoding, a number of $4! = 24$ different encodings are possible. The chosen encoding is responsible for the final CNF representation of each gate and strongly influences the size of the formula. Since a larger CNF typically results in an increased run time of the SAT solver, all possible encodings are evaluated according to the resulting size of their CNF representation for different gate types [FSD06].

The encoding $\eta_{\mathscr{L}_4}$ used in PASSAT is shown in Table 4.2. The upper row denotes the represented values of $\mathscr{L}_4$, whereas each column below shows the corresponding assignment of the Boolean variables c and c^*. This encoding is then used to transform the characteristic function of a gate into CNF. Typically, the use of an encoding signifies an overhead for the size of the CNF representation and slows down the search process compared to the pure Boolean CNF representation.

Example 4.3. Given an AND gate $o = a \cdot b$ represented in $\mathscr{L}_4$. In order to generate the CNF for the AND gate o, two variables are needed for each input and output: a, a^*, b, b^*, o, o^*. Both variables represent the value of the signal. For instance, if $o = 1$ and $o^* = 1$, the value of line o is U. The encoding $\eta_{\mathscr{L}_4}$ is then used to transform the representation of the gate from four-valued logic into Boolean logic using the truth table method and a logic minimizer. The resulting CNF is shown in Table 4.3.

Besides the four-valued logic and the Boolean encoding, PASSAT uses the SAT encoding of D-chains introduced in TEGUS [SBS96]. However, since the D-chains in TEGUS are encoded in Boolean logic, PASSAT introduces an encoding based on $\mathscr{L}_4$ or $\eta_{\mathscr{L}_4}$, respectively.

Table 4.4 CNF size for a 2-input AND gate

	CNF description					
	Boolean			Four-valued		
Constraint	Cls	Lit	∅len	Cls	Lit	∅len
$C^c \equiv (A^c \cdot B^c)$	3	7	2.3	8	23	2.9
$C^f \equiv (A^f \cdot B^f)$	3	7	2.3	8	23	2.9
$C^d \rightarrow (C^c \neq C^f)$	2	6	3.0	5	16	3.2
$C^d \rightarrow (D^d + E^d)$	1	3	3.0	1	3	3.0
Overhead	1.0	1.0	1.0	2.4	2.8	1.1

4.2.2 Hybrid Logic

The use of non-Boolean elements – such as tri-state elements – and unknown values necessitates the application of the four-valued logic $\mathscr{L}_4 = \{0, 1, U, Z\}$ as described above. A Boolean encoding $\eta_{\mathscr{L}_4}$ is applied to transform the multiple-valued ATPG problem into a Boolean problem. As a result, efficient Boolean SAT-based algorithms can be applied. However, the size of the SAT instance increases significantly by applying a Boolean encoding which is shown in Table 4.4.

The number of clauses needed to represent the logical behavior of a 2-input AND gate and its corresponding D-chain encoding are given in columns *Cls*. The number of literals are given in columns *Lit*. The average clause length is given in column *∅len*.

In order to shrink the size of the Boolean formula, the use of a hybrid logic was proposed. Typically, the portion of tri-state elements in an industrial circuit is very small compared to Boolean gates. This also holds for inputs fixed to an unknown state. Both observations can be leveraged to obtain a smaller CNF representation. Based on the sources for the additional values U and Z, a structural analysis is applied to determine which parts of the circuits can be structurally influenced by either U or Z [DEF+08]. These parts are modelled in L_4. Those parts which are not influenced can be modelled in Boolean logic, since it is guaranteed that only Boolean values can be assumed. As a result, the size of the CNF formula can be reduced depending on the portion of gates modelled in Boolean logic. Experiments showed that this portion is typically very large and the size of the CNF formula can be significantly reduced which leads to an accelerated ATPG process.

4.2.3 Improving Compactness

SAT-based ATPG has been shown to be very robust even for large industrial circuits. However, a major problem is that state-of-the-art SAT solvers, e.g. MiniSat [ES04], typically return a complete assignment of all variables in the SAT instance for reason of efficiency. The resulting test pattern is therefore often over-specified. Structural ATPG algorithms such as FAN [FS83] typically assign many signal lines and inputs with don't cares (X) such that the tests can be compacted very well. A small test set is very important in industrial practice and an approach which produces large test sets cannot be used.

In order to overcome this limitation of SAT-based ATPG, a post-processor was proposed in [ED07]. During the generation of the CNF formula, the fault activation as well as the fault propagation path are not known. Therefore, all possible paths are included in the SAT instance. Since a SAT solver returns a complete assignment of all variables, all inputs of the considered circuit part are specified, i.e. no don't care bits are assumed here.

The post-processor uses a path tracing technique to reduce the number of specified input bits. This procedure uses the complete assignment given by the SAT solver and the circuit description to backtrace from the output where the fault effect can be observed towards the inputs. During the backtracing procedure, the necessary assignments for detecting the target fault are determined. All inputs which are not needed for these necessary assignments are re-assigned to the don't care value X. This results in a large number of unspecified bits.

Experimental results showed that the run time overhead of the post-processor is negligible and that the compactness of the generated test patterns is significantly improved.

4.3 Combination with Structural Algorithm

The advantage of SAT-based approaches, e.g. of PASSAT, is the robustness. PASSAT is able to classify more faults than classical structural ATPG approaches due to conflict-based learning and conflict-driven heuristics. In particular, SAT-based approaches are robust for hard-to-test faults. However, these approaches suffer from the overhead of CNF transformation, especially for easy-to-detect faults. A detailed analysis of PASSAT and an industrial structural FAN-based algorithm ("FAN") yields the following observations:

- The FAN-based engine is able to quickly classify a large number of faults, i.e. easy-to-test faults.
- PASSAT typically needs more time for easy-to-test faults due to the circuit-to-CNF transformation.
- The FAN-based engine has serious problems to classify hard-to-detect faults even if more resources are given, i.e. the backtrack limit or run time limit is increased.
- Due to the robustness of the underlying SAT solver, PASSAT often finds a solution for hard-to-detect fault for which FAN could not succeed.

Based on these observations, the following procedure was proposed in [DEF$^+$08]. Figure 4.4a shows the classical flow with a FAN-based algorithm as the main ATPG engine. The combination of the FAN-based algorithm and the SAT-based ATPG algorithm is shown in Fig. 4.4b. Here, the FAN-based algorithm is started at first with restricted resources, e.g. reduced backtrack limit to classify

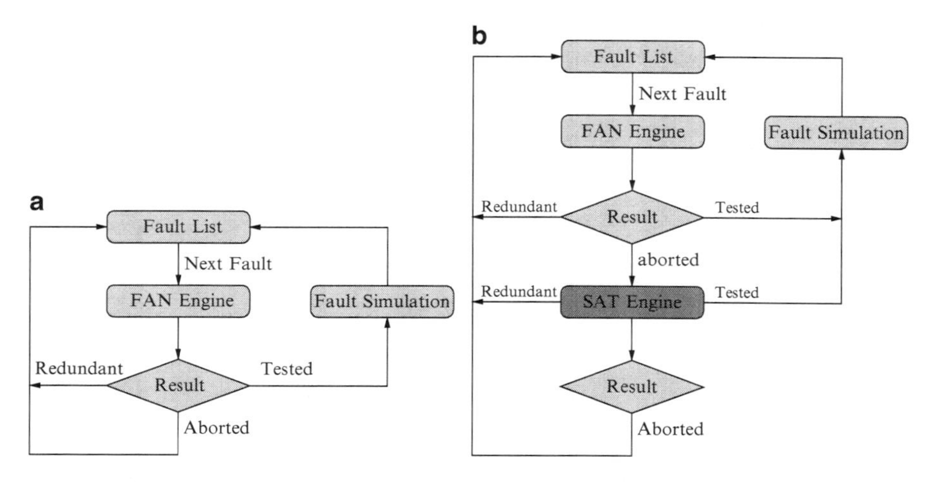

Fig. 4.4 Combination of structural and SAT-based algorithm. (**a**) Classic and (**b**) SAT integration

a particular fault. If a test pattern can be found, the test is fault simulated to detect additional faults and the next fault is processed. If the fault is proven to be redundant, the next fault is targeted by the FAN-based algorithm. If no test pattern can be found ("aborted"), the SAT-based engine is started.

As a result, the FAN-based algorithm quickly classifies a large number of easy-to-detect faults for which the SAT-based engine does not have to be started. Then, the hard-to-detect faults are targeted by the SAT-based algorithm. By this, the advantages of both engines are combined in this flow and the number of aborts produced by structural algorithms can be reduced. However, the advantages of a SAT-based algorithm, in particular the inherent learning feature, are not fully leveraged and much time is spent for unnecessary computations. Many faults are targeted by both approaches. Additionally, a practical problem is that two different ATPG algorithms have to be maintained.

Part II
New SAT Techniques and their Application in ATPG

Chapter 5
Dynamic Clause Activation

Classical ATPG algorithms are very fast and many faults can be classified very quickly. However, these algorithms have problems to cope with hard-to-detect faults whose number is steadily increasing in today's complex designs. This leads to a growing proportion of faults which cannot be classified. At the same time, the quality requirements of the industry demands a high fault coverage for which a large proportion of unclassified faults is disadvantageous. This results in a renewed interest in novel efficient ATPG algorithms which are fast and robust.

As described in the previous chapter, ATPG algorithms based on *Boolean Satisfiability* (SAT) do not work on a circuit structure but on a Boolean formula in *Conjunctive Normal Form* (CNF). Therefore, the ATPG problem is transformed into CNF and a SAT solver is applied to solve the formula. SAT-based ATPG turned out to be an efficient complement to existing classical ATPG algorithms classifying many faults for which structural algorithms determine no solution in reasonable time. Nonetheless, although being very robust in classifying many hard-to-detect faults, SAT-based ATPG algorithms suffer from the overhead for solving easy-to-detect faults which typically represent the majority of all faults. This leads to unacceptable high ATPG run times for large circuits which makes the stand-alone application of SAT-based ATPG in industrial practice unfeasible. In particular, the overhead is caused by the following drawbacks of CNF-based SAT solvers:

- Loss of structural knowledge – Classical ATPG algorithms benefit strongly from the structural knowledge about the ATPG problem. This knowledge is typically lost during the transformation into CNF.
- Transformation into CNF – Although the complexity of the CNF transformation is linear in the number of gates, the transformation time is not negligible. Especially in the ATPG domain where many instances based on the same circuit have to be solved, the transformation time is a significant overhead as reported in [SBS96, MS97].

S. Eggersglüß and R. Drechsler, *High Quality Test Pattern Generation and Boolean Satisfiability*, DOI 10.1007/978-1-4419-9976-4_5,

- Completely specified solution – Due to reasons of efficiency, CNF-based SAT solvers compute a solution where all Boolean variables are specified, although many variables could be assigned with don't cares. This is disadvantageous for efficient test generation as well as for test compaction techniques.

This chapter presents the SAT technique *Dynamic Clause Activation* (DCA) [ED11b]. Using DCA, the SAT solver works on a subset of the original problem instance which is extended dynamically. This procedure has the following advantages:

- Exploitation of structural knowledge – Structural knowledge can be used during the search process due to retaining the gate connectivity information.
- Transformation into CNF – The CNF for the circuit is created only once. The required clauses are dynamically activated and by this the overhead of creating a complete SAT instance for each fault is significantly reduced.
- Implicit modeling of *Observability Don't Cares* (ODCs) – Due to the DCA technique, ODCs are modeled implicitly. By this, SAT core techniques, e.g. *Boolean Constraint Propagation* (BCP) and conflict analysis, do not have to be modified and retain their efficiency. Additionally, the generated tests contain an increased number of unspecified bits.

The modeling of ODCs or the utilization of structural knowledge is not new as described in the previous chapter. However, DCA is the first technique that permits the efficient combination. Core techniques such as fast BCP and conflict-driven learning are not altered. The application of DCA in SAT-based ATPG results in a significant speed-up. At the same time, the high level of robustness of SAT-based ATPG can be retained. Furthermore, the flexibility of this technique allows for a dynamically switching of the search mode, i.e. classical SAT solving without DCA can be activated during the search if this turns out to be advantageous (*emulation*).

This chapter is organized as follows: Sect. 5.1 presents the overall framework of a SAT engine using DCA. In this context, the changes to the classical DLL algorithm necessary for the integration of the DCA technique are presented. The activation methodology is given in Sect. 5.2. Section 5.3 shows improved activation rules and Sect. 5.4 describes the modeling of implicit observability don't cares. The emulation of a classical SAT solver is given in Sect. 5.5. Section 5.6 presents the SAT formulation for test generation needed for using DCA. Experimental results are given in Sect. 5.7. Section 5.8 summarizes this chapter.

5.1 Overall Framework for Dynamic Clause Activation

A large part of the run time needed by a SAT-based algorithm stems from the CNF generation as reported in [SBS96, MS97]. The transformation of the ATPG problem into a SAT problem is a significant overhead, in particular for easy-to-detect faults. Furthermore, easy-to-detect faults often represent the majority of faults. The SAT

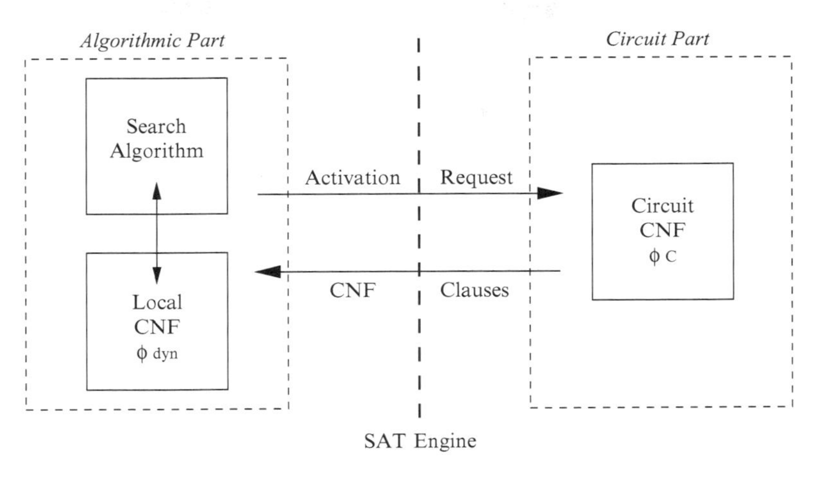

Fig. 5.1 Division of the proposed SAT engine

instance Φ_{test}^{F} for test generation for fault F consists of two parts: the functionality of $\mathscr{C}$ (represented by $\Phi_{\mathscr{C}}$) and the fault-specific constraints Φ_F. Fault-specific constraints are those constraints which are added to the circuit CNF in order to generate a test for a specific fault:

$$\Phi_{\text{test}}^{F} = \Phi_{\mathscr{C}} \cdot \Phi_F$$

A large number of SAT instances have to be built for each circuit for ATPG. Since the same parts of the circuit are considered very often, large parts of the CNF can potentially be shared between faults and the CNF transformation time can be reduced [FWD07]. However, determining which part of the CNF can be shared with a subsequent fault is very complex and costly. To complicate matters further, fault dropping is typically enabled in industrial practice during test generation. As a result, subsequent faults usually only share a small part or even no part.

The DCA technique presented in this book bypasses this problem by building the CNF for the circuit only once. Then, the necessary parts of the circuit are dynamically activated and deactivated. In contrast to common SAT solvers, the approach does not work on a static SAT instance Φ_{test}^{F}, but on a dynamic SAT instance named Φ_{dyn} which is extended during the search process. Initially, for each fault, $\Phi_{\text{dyn}} = \Phi_F$ holds. During the search process, Φ_{dyn} is extended dynamically with clauses from $\Phi_{\mathscr{C}}$.

In order to allow for a dynamic extension of the SAT instance, the proposed SAT engine using DCA is divided into two parts: a *circuit part* and an *algorithmic part*. This is illustrated in Fig. 5.1. However, both parts are tightly integrated. The circuit part contains the complete CNF of the circuit ($\Phi_{\mathscr{C}}$) and serves as a database, whereas the algorithmic part contains the search algorithm working on a local CNF

Algorithm 3 Deduce()

```
1: Literal l; /* l stores newly found implication */
2: while STATUS = Find_next_implication(l) do
3:     if STATUS == CONFLICT then
4:         return CONFLICT;
5:     else
6:         Assign(l);
7:         Activate(l); /* Clauses associated with l are activated */
8:     end if
9: end while
10: return NO_CONFLICT;
```

Algorithm 4 Decide()

```
1: Literal l = Pick_activated_branch_literal(); /* Returns ∅ if no literal can be found */
2: if l == ∅ then
3:     return FALSE;
4: else
5:     Assign(l);
6:     Activate(l); /* Clauses associated with l are activated */
7:     return TRUE;
8: end if
```

Φ_{dyn} representing the set of activated clauses. The search algorithm consists of the DLL algorithm and state-of-the-art SAT techniques such as conflict analysis and fast BCP and is only executed on Φ_{dyn}. In fact, each SAT solver can be taken as basis for the algorithmic part and extended to fit in a framework using DCA.

In the following, the integration of the DCA technique into the DLL algorithm is described. The DLL algorithm is given in Algorithm 2 (see Sect. 3.2.1). The only changes have been made in the sub-routines *deduce()* and *decide()* which are given in Algorithms 3 and 4, respectively. The *decide()* function is changed in order to carry out decisions on variables in activated clauses only. This is given by the function *Pick_activated_branch_literal()* (line 1).

Whenever a new assignment is made – either decision or implication – an activation request is sent (line 7 in Algorithm 3 and line 6 in Algorithm 4) before the new assignment is propagated. The activation request takes care that the corresponding part of the circuit is activated and Φ_{dyn} is dynamically extended during the search process. Each request has to be sent before doing BCP for this assignment. Otherwise, a potential conflict can be missed and the integrity of the search process is violated. In the following, a clause ω is called activated if ω is contained in Φ_{dyn}.

Using this procedure, SAT (testable) is determined if all variables contained in activated clauses are assigned and no conflict exists (analogously to the break condition of common SAT solvers). That is, when all activated clauses are satisfied. In contrast, if a conflict in Φ_{dyn} cannot be resolved, then the SAT instance is UNSAT, i.e. the fault is redundant, because any extension of Φ_{dyn} would also result in

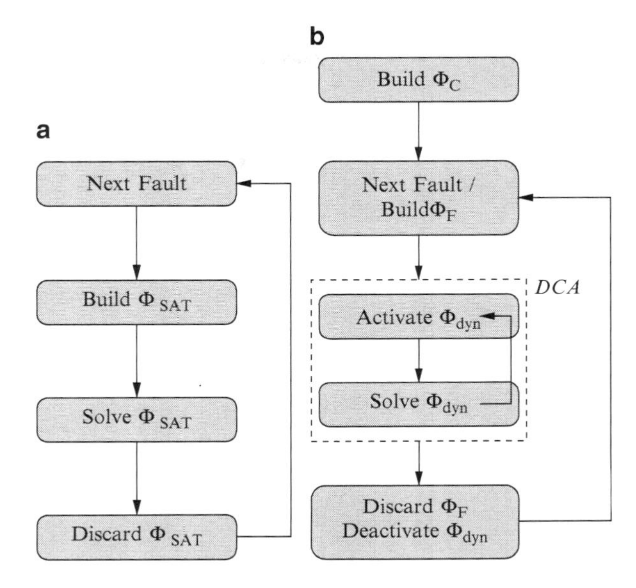

Fig. 5.2 SAT-based ATPG flow. (**a**) Classic and (**b**) dynamic

UNSAT. By this, redundant faults can be classified much faster when the reason of the untestability is locally bounded. In this case, small parts of the complete CNF only have to be activated.

After solving one formula, Φ_{dyn} is cleared. All clauses are deactivated. The classical test generation flow as well as the alternated flow incorporating DCA is shown in Fig. 5.2a, b, respectively. The time-consuming step of circuit-to-CNF transformation is done only once instead of for each fault. This step is replaced by the activation methodology which can be efficiently implemented using pointers to clauses. Only Φ_F has to be extracted for each run. The size of the fault-specific constraints Φ_F is typically very small in relation to the size of the circuit CNF. However, holding the complete circuit CNF in the memory during the ATPG process imposes an increased memory consumption. Section 5.6 shows the detailed SAT formulation for an ATPG problem using DCA.

Crucial for an efficient application of this technique is the correct application of the activation methodology, which is described next.

5.2 Efficient Activation Methodology

The activation methodology has to consider the following properties:

- Consistent SAT instance – Guaranteeing that the dynamically activated SAT instance is consistent with the classically formulated SAT instance is important to avoid overhead in processing the CNF and ensure the correctness of the calculation.

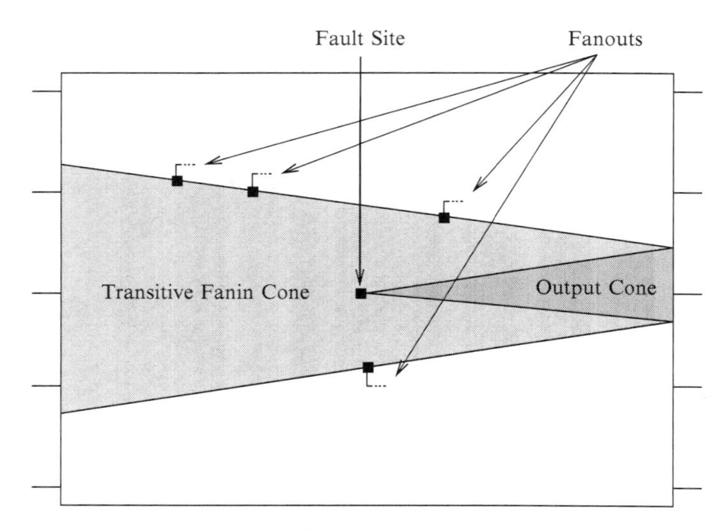

Fig. 5.3 Transitive fanin cone with fanouts

- Efficiency of the search process – The efficiency of a SAT solver is highly susceptible for changes. Consequently, the interference of the activation methodology into the search process has to be as small as possible.

5.2.1 Consistent SAT Instance

A conservative approach to decide which clauses should be activated when variable x is assigned, is to activate all clauses containing x. Since these are the only clauses in which an implication can be found or a conflict can be detected, respectively, caused by the assignment of x. However, this is disadvantageous because also clauses from other parts of the circuit – which do not have to be considered for a specific fault – would be activated. In the worst case, the complete circuit is assigned which is a significant overhead for test generation. The reason for this is demonstrated in Fig. 5.3. If at least one fanout branch b of a fanout stem contained in the transitive fanin cone is outside the transitive fanin cone, i.e. outside the considered part of the circuit, the circuit part associated with branch b is automatically included in the activation procedure. As a result, this approach does not yield a SAT instance which is consistent with the classically formulated SAT instance but a significantly larger one. This obviously imposes overhead for the search process.

Instead, a different activation methodology is proposed. As explained in Sect. 4.1.1, the ATPG problem can be solved by justification. Justification corresponds to the search for an input assignment of a gate which results to the desired output value. The proposed activation methodology is also influenced by the idea of justification. If a value v is assigned to a variable x, those clauses should be activated which are needed to justify v. Since a variable is associated with a connection s and, consequently, s is the outgoing line of one gate g, the set of clauses needed for justification is Φ_g, i.e. the clauses of the preceding gate. In the following, a variable x is denoted as activated if an activation request has been sent as a consequence of the assignment of x.

Only the clauses of the preceding gates are activated. The clauses of the succeeding gates remain deactivated (if these clauses haven't been activated before). This ensures that only clauses of the transitive fanin cone become activated. The activation methodology has an inherent stopping criterion. Because inputs do not have preceding gates, i.e. for input i, $\Phi_i = \emptyset$ holds, they do not need to be justified and no clauses are activated. The proposed activation methodology is very similar to the classical SAT instance generation. Both use information of the graph structure. Therefore, if the complete SAT instance is activated, it corresponds to the classically formulated SAT instance. More formally, given a connection $s = g \times h$ and a Boolean variable x_g associated with s, then the clauses contained in Φ_g are activated if x_g is assigned to either 0 or 1.

Given the classically formulated SAT instance $\Phi^F_{\text{test}} = \Phi_F \cdot \Phi_{\mathscr{C}}$ and the initial SAT instance for DCA $\Phi_{\text{dyn}_0} = \Phi_F$ (see Sect. 5.1), the SAT instance Φ_{dyn_0} is extended during the search process resulting in the SAT instances $\Phi_{\text{dyn}_1}, \ldots, \Phi_{\text{dyn}_n}$. Here, $\Phi_{\text{dyn}_n} = \Phi^F_{\text{test}}$ is the resulting SAT instance when no more activation requests are sent.

Nevertheless, the proposed activation methodology relies on the utilization of structural knowledge. A data structure which provides this knowledge is shown next.

5.2.2 *Structural Watch List*

The *Structural Watch List* (SWL) – a new data structure which is located in the circuit part – is proposed for the use of the activation methodology. The SWL provides structural information and, at the same time, allows for an efficient integration of the activation methodology.

The implementation of the SWL is strongly influenced by the watch lists used in modern SAT solvers for fast BCP (see Sect. 3.3.1). The SWL is implemented as a doubly nested list. That means, each entry in the SWL represents a list of clauses which should be activated when a specific activation request is sent, i.e. if a specific variable was assigned. Therefore, each variable x in the circuit is associated with an entry in the SWL. If a value is assigned to x during the search process, then those clauses stored in the associated list entry are activated, i.e. added to Φ_{dyn}.[1]

---

[1]The dynamic extension of Φ_{dyn} can be efficiently implemented by adding references/pointers of the activated clauses to the watch lists used for fast BCP. By this, the SAT algorithm does not have to be changed and keeps its effectiveness.

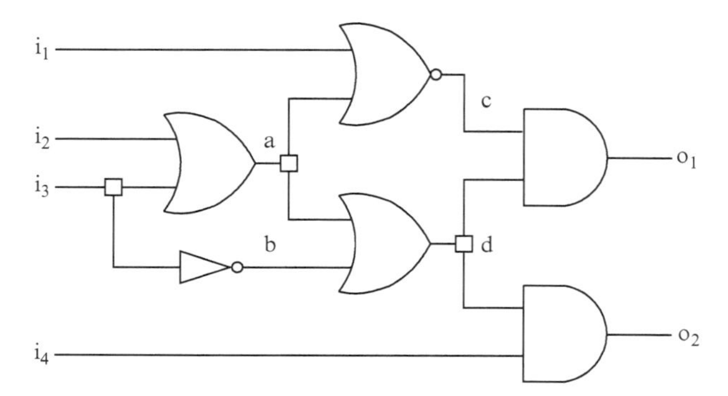

Fig. 5.4 Example circuit $\mathscr{C}$ for dynamic clause activation

Table 5.1 CNF and SWL entries for the example circuit $\mathscr{C}$ in Fig. 5.4

Id	Entry	Clause	Id	Entry	Clause
ω_1	o_1	$(o_1+\overline{c}+\overline{d})$	ω_{10}	a	$(\overline{a}+i_2+i_3)$
ω_2	o_1	$(\overline{o}_1+c)$	ω_{11}	a	$(a+\overline{i}_2)$
ω_3	o_1	$(\overline{o}_1+d)$	ω_{12}	a	$(a+\overline{i}_3)$
ω_4	c	$(c+i_1+a)$	ω_{13}	b	$(\overline{b}+\overline{i}_3)$
ω_5	c	$(\overline{c}+\overline{i}_1)$	ω_{14}	b	$(b+i_3)$
ω_6	c	$(\overline{c}+\overline{a})$	ω_{15}	o_2	$(o_2+\overline{d}+\overline{i}_4)$
ω_7	d	$(\overline{d}+a+b)$	ω_{16}	o_2	$(\overline{o}_2+d)$
ω_8	d	$(d+\overline{a})$	ω_{17}	o_2	$(\overline{o}_2+i_4)$
ω_9	d	$(d+\overline{b})$			

As described in Sect. 3.4, a variable x_g is associated with a connection, while the clauses Φ_g represent the function of the preceding gate g. All clauses contained in Φ_g are therefore registered at the SWL entry of variable x_g. As a result, if g is assigned a value, all clauses of Φ_g will be activated before doing BCP for the assignment of x_g. By this, some structural information about the directed graph structure of the circuit is included. This allows for the justification of the value assigned to x_g. Note that D-chain constraints (as introduced in Sect. 4.1.2) can be handled in a similar way. All clauses representing D-chain constraints are registered at the SWL entry of the corresponding d-variable. Therefore, if the d-variable is assigned, the corresponding clauses are activated. The next example demonstrates the construction of the SWL for a specific circuit.

Example 5.1. Consider the example circuit $\mathscr{C}$ presented in Fig. 5.4. Ten variables are needed to represent the circuit in CNF. Consequently, the SWL has ten entries; each for one variable. Table 5.1 shows the CNF or the clauses, respectively, representing the circuit's function. In column *Id*, the name of the clause is given and column *Clause* presents the clause itself. Column *Entry* shows at which variable the clause is registered in the SWL. Figure 5.5 gives the schematic view of the SWL.

Next, the progress of the search process using DCA is shown. The implication graph of an example search process is given in Fig. 5.6. Assume that the target is

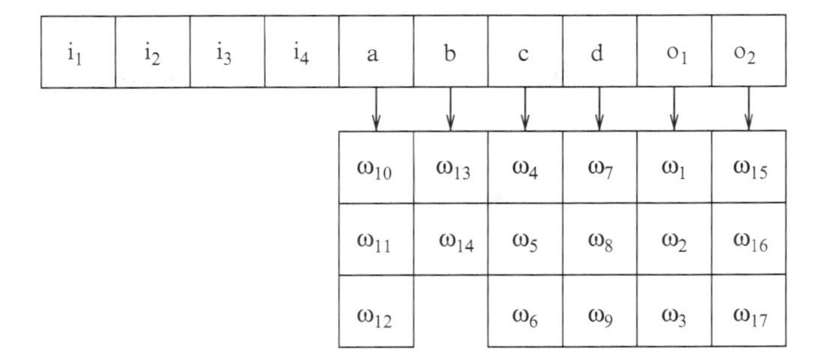

Fig. 5.5 Schematic view of the SWL

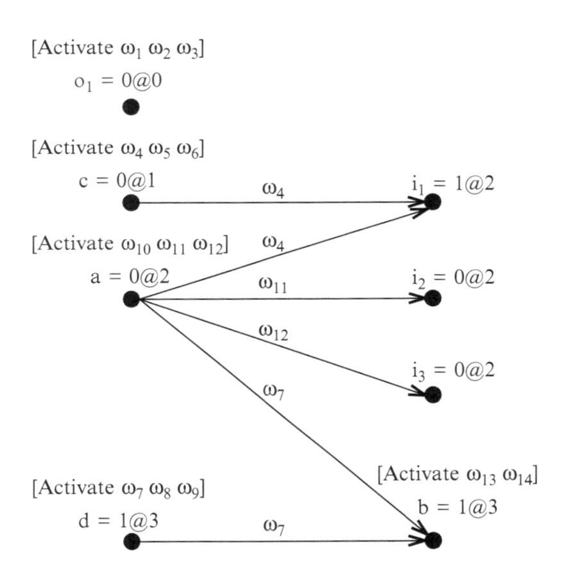

Fig. 5.6 Example implication graph for search process using DCA

to justify the value $o_1 = 0$. Therefore, the first assignment is $o_1 = 0$ followed by the activation of the clauses ω_1, ω_2 and ω_3, which represent the function of the preceding AND gate. Decisions are carried out on those variables which occur in activated clauses. The next decision is $c = 0$ which causes the activation of ω_4, ω_5 and ω_6. The first implications are possible with the decision $a = 0$ in decision level 2.

The activated clauses ω_4, ω_{11} and ω_{12} lead to the implications $i_1 = 1, i_2 = 0$ and $i_3 = 0$. Since these variables are associated with inputs, no activation is performed. The SWL entry is empty. The last unassigned variable contained in activated clauses is d. The decision $d = 1$ leads to the implication $b = 1$, because ω_7 has been activated. This in turn leads to the activation of ω_{13} and ω_{14}. Both clauses are satisfied under the current assignment. As a result, all variables contained in activated clauses are satisfied and no conflict exists. Therefore, the search process yields the result SAT and the satisfying assignment:

$$i_1 = 1, i_2 = 0, i_3 = 0, a = 0, b = 1, c = 0, d = 1, o_1 = 1$$

Those clauses corresponding to gates outside the fanin cone of o_1 were not activated and the associated variables o_2, i_4 were not assigned.

5.3 Literal-Based Activation

The variable-based activation methodology can be improved by literal-based activation. Since the activation of clauses is done gate by gate, the CNF of a gate g is examined in more detail to motivate the literal-based activation. This set of clauses can be divided into two parts:

$$\Phi_g = \Phi_g^0 \cdot \Phi_g^1$$

The CNF Φ_g^0 is satisfied if $g = 0$ and Φ_g^1 is satisfied if $g = 1$. This is illustrated by the CNF for a NAND gate g in the following example. The application for other gate types is straightforward.

Example 5.2. Consider the NAND gate $g = \overline{a \cdot b}$. The following three clauses are used to represent the functionality in CNF:

$$\begin{aligned} \omega_1 &= (\overline{g} + \overline{a} + \overline{b}) \\ \omega_2 &= (g + a) \\ \omega_3 &= (g + b) \end{aligned}$$

Under the assignment $g = 0$, $\omega_1 = 1$ holds, i.e. ω_1 is satisfied, while $\omega_2 = 1$ and $\omega_3 = 1$ hold under the assignment $g = 1$. Φ_g^0 and Φ_g^1 are composed as follows:

$$\begin{aligned} \Phi_g^0 &= \{\omega_1\} \\ \Phi_g^1 &= \{\omega_2, \omega_3\} \end{aligned}$$

This clause division can be leveraged in the DCA activation rules. An assignment $g = v$ where $v \in \{0, 1\}$ results in a satisfied set of clauses Φ_g^v. Since the target is to satisfy all clauses of a gate g, only those clauses have to be activated which are unsatisfied under a current assignment of g. This correlates directly to the justification rules used in structural ATPG. If the controlling value (the inverted controlling value in case of an inverting gate) is assumed on the output of gate g, it is sufficient that only one input assumes the controlling value of g. The other inputs do not influence the value on the output and are *don't care*. Again, this is demonstrated by the CNF of a NAND gate.

Table 5.2 SWL entries for literal-based activation for circuit $\mathscr{C}$ in Fig. 5.4

Id	Entry	Clause	Id	Entry	Clause
ω_1	$\overline{o}_1$	$(o_1+\overline{c}+\overline{d})$	ω_{10}	a	$(\overline{a}+i_2+i_3)$
ω_2	o_1	$(\overline{o}_1+c)$	ω_{11}	$\overline{a}$	$(a+\overline{i}_2)$
ω_3	o_1	$(\overline{o}_1+d)$	ω_{12}	$\overline{a}$	$(a+\overline{i}_3)$
ω_4	$\overline{c}$	$(c+i_1+a)$	ω_{13}	b	$(\overline{b}+\overline{i}_3)$
ω_5	c	$(\overline{c}+\overline{i}_1)$	ω_{14}	$\overline{b}$	$(b+i_3)$
ω_6	c	$(\overline{c}+\overline{a})$	ω_{15}	$\overline{o}_2$	$(o_2+\overline{d}+\overline{i}_4)$
ω_7	d	$(\overline{d}+a+b)$	ω_{16}	o_2	$(\overline{o}_2+d)$
ω_8	$\overline{d}$	$(d+\overline{a})$	ω_{17}	o_2	$(\overline{o}_2+i_4)$
ω_9	$\overline{d}$	$(d+\overline{b})$			

Example 5.3. Consider the CNF of a NAND gate g as given in Example 5.2. The controlling value of a NAND gate is 0. Note that a NAND gate is an inverting gate. Therefore, if the value 1 has to be justified on the output of g, only one input has to assume the controlling value 0. This is represented in CNF by the clause ω_1 contained in Φ_g^0. The literal $\overline{g}$ in clause ω_1 is unsatisfied. In order to satisfy ω_1, either $\overline{a}$ or $\overline{b}$ has to be satisfied. In contrast, if the value 0 has to be justified, the set of clauses Φ_g^1, i.e. the clauses ω_1 and ω_2, are relevant since they represent the implications.

The activation procedure is altered as follows. Given a connection $s = g \times h$, the output of gate g and a Boolean variable x_g associated with s, then

$$(x_g = 0) \rightarrow \text{Activate}(\Phi_g^1)$$
$$(x_g = 1) \rightarrow \text{Activate}(\Phi_g^0)$$

By this, only a subset of clauses has to be activated since some of the gate clauses are known to be satisfied. The procedure is complete since the remaining clauses will be activated when the value of x_g is flipped in the further search.

The SWL has to be altered for the implementation of the literal-based activation. Instead of one entry for each variable x in the circuit, two entries are necessary: one for the positive form and one for the negative form (similar to the *two-literal watch scheme* used for fast BCP). In other words, one entry is assigned to each literal of x. Each entry contains those clauses which should be activated if the variable is assigned with the corresponding value. This is Φ_g^1 for literal $\overline{g}$ and Φ_g^0 for literal g.

Consider again the example circuit shown in Fig. 5.4. The SWL entries using literal-based activation are shown in Table 5.2. Figure 5.7 gives the schematic view of the SWL. Note that the empty lists of the inputs are not shown in this figure.

The advantage of the literal-based activation compared to variable-based activation is that typically less clauses are activated. This is beneficial for the BCP procedure since satisfied clauses are generally kept in the watch lists in modern SAT solvers because it causes too much overhead to remove them.

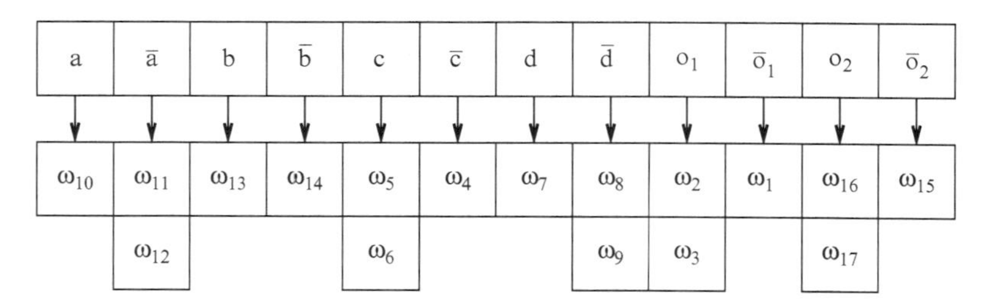

Fig. 5.7 Schematic view of the SWL for literal-based activation

5.4 Implicit Observability Don't Cares

The described activation methodology can be enhanced by the implicit modeling of ODCs. Contrary to structural ATPG algorithms, SAT solvers cannot explicitly assign a variable with a don't care value. Their efficiency is based in part on the fact that their search is defined over Boolean variables. The activation methodology modeling ODCs is motivated by the following property of a circuit-oriented problem: if the controlling value of a gate g is assigned to an input of g, the remaining inputs of g do not influence the output value of g anymore. The value of the other inputs can be considered as don't care. Furthermore, if the problem is to justify certain values in the circuit, these don't care inputs do not influence the search at all. A variable which does not influence the outcome of the search process can be denoted as ODC as defined in [SVD08].

Previous SAT-based approaches applied in other circuit-oriented domains than ATPG enhanced the SAT algorithm by structural information and explicitly declare variables or signal lines as don't care. The procedure to include ODCs presented here differs from previous approaches since ODCs are implicitly included in the search process. The advantage of an implicit inclusion is that the search algorithm and the efficient SAT techniques which are very susceptible to modifications do not have to be changed.

An *Implicit Observability Don't Care* (IODC) is a variable that remains unconstrained by not activating the clauses needed for justification. That means, if a clause ω is already satisfied but not all variables contained in ω are assigned, the assignment of these variables is not necessary. As a result, no activation request has to be sent for them and these variables remain unconstrained, although they do not correspond to inputs. In order to avoid that a value is assigned to an IODC variable and, consequently, becomes constrained by an activation request, the break condition of the solving process has to be changed. SAT is determined if all activated clauses are satisfied and no activation requests are to be sent. All unassigned variables occurring in activated clauses are IODCs. The following example shows the handling of IODCs using literal-based activation.

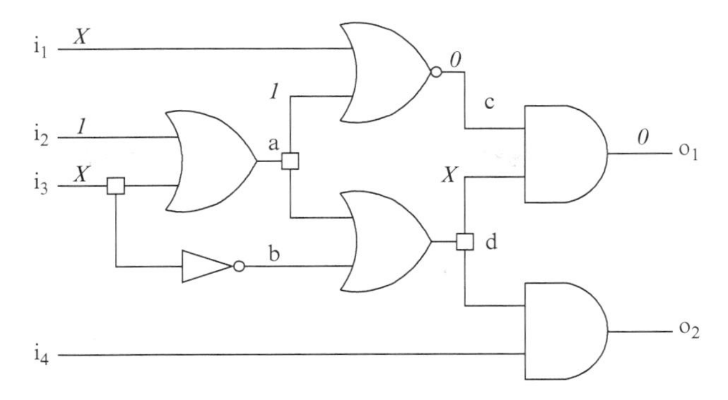

Fig. 5.8 Example circuit C with IODCs

Example 5.4. Figure 5.8 shows the example circuit used in Example 5.1. Again, assume that the target is to justify the value 0 on output o_1. According to the generated SWL (presented in Table 5.2), ω_1 is activated. In order to satisfy ω_1, variable c is assigned with 0 and ω_4 is activated. Then, the decision $a = 1$ is carried out. As a result, ω_{10} is activated. The next decision is $i_2 = 1$. Now, all activated clauses $(\omega_1, \omega_{10}, \omega_4)$ are satisfied and a solution is found, although not all variables in activated clauses are assigned:

$$i_2 = 1, a = 1, c = 0, o_1 = 0$$

The variables i_1, i_3, d are also contained in $\omega_1, \omega_{10}, \omega_4$ but they remain unconstrained and consequently are IODCs.

A solution can be found more quickly since large parts of the search space might be pruned by the use of IODCs. However, the disadvantage is that the break condition has to be altered. If many clauses are activated, the continuous check whether all clauses are satisfied generates a large overhead.[2] Therefore, the use of a clause-based *J-frontier* is proposed to diminish the overhead.

5.4.1 Clause-Based J-Frontier

Since the number of activated clauses grows the longer the search process endures, the overhead of checking for satisfiability increases simultaneously if IODCs are considered. This section describes the usage of a J-frontier. Using a J-frontier to determine satisfiability has the advantage that the number of clauses which has

[2]In fact, this was done in earlier SAT solvers. Due to reasons of efficiency, modern SAT solvers check whether all variables are assigned and no conflict exists.

to be checked remains small and the overhead is limited. The use of a J-frontier is not new in the field of ATPG. A J-frontier is a classical concept to keep track of unjustified lines [ABF90]. The SAT-based approach CGRASP [GSM99] uses a J-frontier denoting the set of variables requiring justification. Conditions were formulated to determine which variables are contained in the set.

However, the modeling of the J-frontier presented in this book is different. The J-frontier is modeled as a clause stack which is tightly coupled with the DCA procedure. In the following, this stack is called *J-stack* and denoted by $\mathscr{J}$. Whenever a clause is activated, the clause is pushed on $\mathscr{J}$. If all clauses on $\mathscr{J}$ are satisfied, the SAT instance is satisfiable. The difference to the pure activation procedure described above is that, when backtracking is performed during the search process, clauses are popped from $\mathscr{J}$.

Each clause pushed on $\mathscr{J}$ was activated due to an activation request or a specific assignment, respectively. If the assignment was undone, the justification of this value is not necessary anymore. As a result, those clauses which were activated to justify this value can be popped from the stack. Alternatively, these clauses could be deactivated. However, deactivating single clauses is a severe intervention into the data structures of the SAT solver and turned out to be very inefficient. Instead, a stack can be efficiently implemented as a dynamically growing list of clause references. Only push and pop operations have to be performed on this stack. The push and pop operations fit very well in the actual handling of assignments and backtracks of modern SAT solvers. In the following example, the use of the J-stack is demonstrated.

Example 5.5. Consider the circuit shown in Fig. 5.9 which was already used in Example 5.1 and Example 5.4. Again, the target is to justify the value 0 on output o_1. The algorithm proceeds as follows (consider the SWL in Table 5.2):

1. Assign $o_1 = 0$
2. Activate ω_1
3. Push ω_1 on $\mathscr{J} \rightarrow \mathscr{J} = \{\omega_1\}$
4. Assign $d = 0$
5. Activate ω_8, ω_9
6. Push ω_8, ω_9 on $\mathscr{J} \rightarrow \mathscr{J} = \{\omega_1, \omega_8, \omega_9\}$
7. Imply $a = 0, b = 0$
8. Activate $\omega_{11}, \omega_{12}, \omega_{13}, \omega_{14}$
9. Push $\omega_{11}, \omega_{12}, \omega_{13}, \omega_{14}$ on $\mathscr{J} \rightarrow \mathscr{J} = \{\omega_1, \omega_8, \omega_9, \omega_{11}, \omega_{12}, \omega_{13}, \omega_{14}\}$
10. Imply $i_2 = 0, i_3 = 0, i_3 = 1$
11. κ at i_3.

The conflicting assignment is shown in Fig. 5.9a. The J-stack consists of the following clauses:

$$\mathscr{J} = \{\omega_1, \omega_8, \omega_9, \omega_{11}, \omega_{12}, \omega_{13}, \omega_{14}\}$$

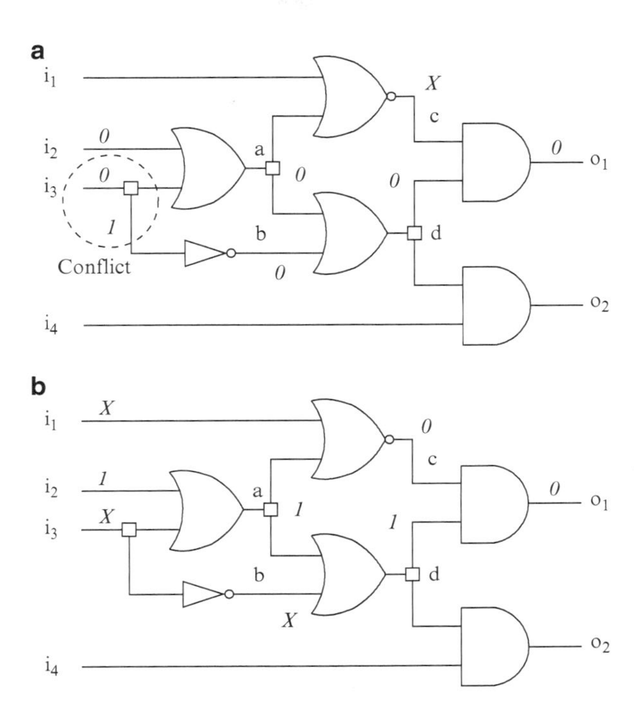

Fig. 5.9 Example – J-stack. (**a**) Conflicting assignment and (**b**) valid assignment

Next, backtracking is performed and assignments are undone. The value of variable d is flipped and $\mathscr{J}$ is updated:

1. Remove assignments of i_3, i_2, a, b
2. Pop $\omega_{14}, \omega_{13}, \omega_{12}, \omega_{11}, \omega_9, \omega_8$ from $\mathscr{J} \rightarrow \mathscr{J} = \{\omega_1\}$
3. Flip d, $d = 1$
4. Activate ω_7
5. Push ω_7 on $\mathscr{J} \rightarrow \mathscr{J} = \{\omega_1, \omega_7\}$
6. Assign $c = 0$
7. Activate ω_4
8. Push ω_4 on $\mathscr{J} \rightarrow \mathscr{J} = \{\omega_1, \omega_7, \omega_4\}$
9. Assign $a = 1$
10. Activate ω_{10}
11. Push ω_{10} on $\mathscr{J} \rightarrow \mathscr{J} = \{\omega_1, \omega_7, \omega_4, \omega_{10}\}$
12. Assign $i_2 = 1$
13. SAT detected.

Under the current assignment

$$o_1 = 0, d = 1, c = 0, a = 1, i_2 = 1$$

the SAT instance is satisfiable since all clauses on the J-stack

$$\mathscr{J} = \{\omega_1 = (o_1 + \overline{c} + \overline{d}), \omega_7 = (\overline{d} + a + b), \omega_4 = (c + i_1 + a), \omega_{10} = (\overline{a} + i_2 + i_3)\}$$

are satisfied. Figure 5.9b shows the valid assignment of the circuit in order to produce $o_1 = 0$.

Note that if an assignment is repeated, i.e. the previous assignment has been done before a backtrack, no activation is performed, because the clauses were already activated. However in this case, the J-stack has to be updated before checking the break condition. The representation of the J-stack can be improved by only pushing clauses on $\mathscr{J}$ which have more than two literals. This is possible, because each binary clause is activated as a result of an assignment. Then, either a conflict is detected or an implication is found directly. Therefore, binary clauses do not have to be checked explicitly. By this, the effort in maintaining the J-stack and checking the break condition is decreased significantly.

5.4.2 SCOAP-Based Decision Heuristic

The decision heuristic of a structural ATPG algorithm is typically influenced by testability measurements, e.g. SCOAP [GT80]. An overview on testability measurements and how to derive them can be found in [BA00]. SCOAP values are computed as a pre-process. The observability value of a line indicates the difficulty of propagating a fault on this line to an output. In contrast, the controllability value indicates the difficulty of justifying a value on this line. Two different controllability values are computed for each line: *control-0* and *control-1*. Each of them denotes the difficulty to justify the corresponding value.

The common decision heuristic of a SAT solver is conflict-driven and do not use any structural information. In [TELD09], the decision heuristic of a SAT solver is influenced by SCOAP values. An initial variable order is created based on the SCOAP values of variables. This order is then dynamically changed by the conflict-driven heuristic of the SAT solver.

However, SCOAP values can also be leveraged using DCA and the J-stack. The proposed decision heuristic makes use of previously computed SCOAP values. The SCOAP values are stored such that each variable which corresponds to a line is assigned the control-0 and the control-1 value while the observability value is assigned to d-variables (see Sect. 4.1.2). The division into d-variables and normal variables has the reason that d-variables are responsible for the propagation path. Here, the observability value is more important.

Algorithm 5 shows the altered pseudo-code of the decision procedure. At first, an unsatisfied clause ω is chosen from $\mathscr{J}$. Then, a literal is chosen from ω in order to satisfy the clause. Here, the SCOAP values of all literals are compared. Which

Algorithm 5 Decide()

```
1: Clause ω = Pick_unsatisfied_clause(𝒥); /* Returns ∅ if all clauses on 𝒥 are satisfied */
2: if ω == ∅ then
3:     return FALSE;
4: else
5:     Literal l = Pick_unassigned_literal_with_minimal_value(ω);
6:     Assign(l);
7:     Activate(l); /* Clauses associated with l are activated */
8:     return TRUE;
9: end if
```

SCOAP value is chosen for comparison is influenced by the type of variable and the polarity of the literal:

- The observability value is chosen for all d-variables independently from the polarity of the literal.
- The control-0 value is chosen for literals (non-d-variables) with negative polarity ($\bar{l}$).
- The control-1 value is chosen for literals (non-d-variables) with positive polarity (l).

Applying a SCOAP-based heuristic guides the search that a solution can be potentially found faster. The advantage of exploiting SCOAP values in combination with DCA is that IODCs can be applied more effectively in contrast to common SAT-based ATPG, because not all variables have to be assigned.

5.5 Classical SAT Solver Emulation

The use of DCA with J-stack and SCOAP-based decision heuristic signifies a renunciation of the classical conflict-driven techniques/heuristics of a SAT solver. SAT-based ATPG is a complement to structural ATPG being very robust for hard-to-detect faults. The application of DCA results in a significant speed-up for easy-to-detect faults. At the same time, the robustness decreases slightly due to the decreasing influence of the conflict-driven heuristics.[3] However, due to the flexibility of the DCA technique, a classical SAT solver can be emulated to retain a high robustness.

The following procedure which is illustrated in Fig. 5.10 is proposed for emulation. The search for a solution is divided into two parts: the DCA part and the emulation part. At first, the search process starts as described above with the initial SAT instance Φ_F using DCA with J-stack and a SCOAP-based decision heuristic.

[3]Nonetheless, the robustness of SAT-based ATPG using DCA is very high compared to structural ATPG as the experiments will show.

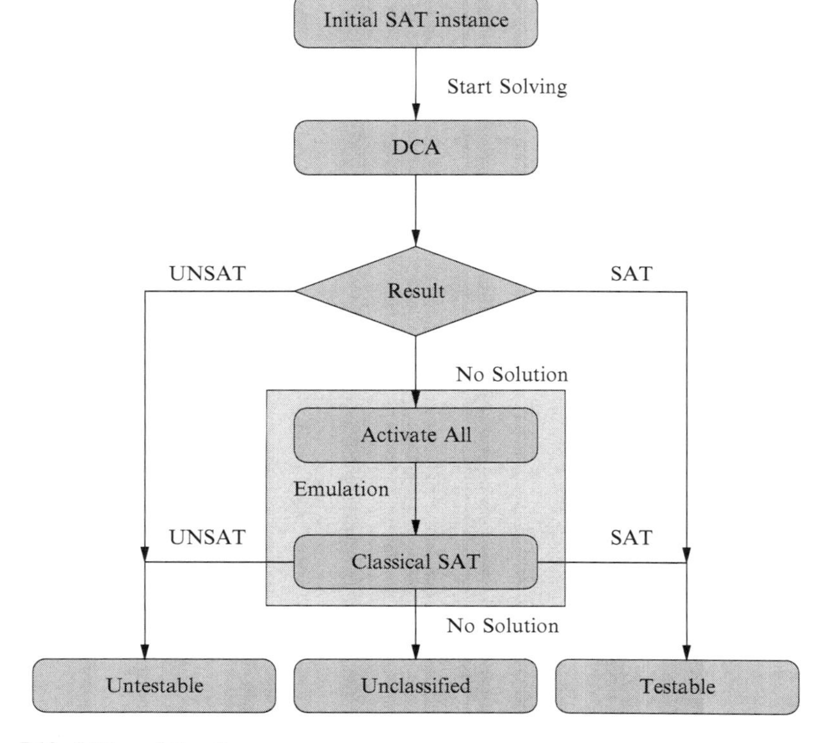

Fig. 5.10 SAT emulation flow

If no solution can be found in a specific time interval, the SAT solver emulation is started. All clauses that are not activated but can be potentially activated for the current fault are added to Φ_{dyn} at once. That is, they become activated. Furthermore, the J-stack is disabled and the conflict-driven decision heuristic is used. By this, the SAT solver acts as a classical SAT solver. Because two different engines are merged into one, this approach is also called *hybrid* approach.

This procedure slightly resembles the combination of a structural and SAT-based ATPG algorithm proposed in [DEF$^+$08]. There, a structural ATPG algorithm is used to quickly classify a large number of faults. If no solution can be found in a short time interval, the SAT-based algorithm is started to find a solution. However, the hybrid approach acts differently since only one engine is used and the integration is more tight. Parts of the search tree which were already traversed in the DCA part do not have to be traversed in the emulation part again. Furthermore, learned information, i.e. conflict clauses, generated in the DCA part is also available in the emulation part.

5.6 SAT Formulation for Test Generation Using DCA

The SAT instance creation for DCA differs slightly from the SAT instance generation for classical SAT-based ATPG which was described in Sect. 4.1. Therefore, the SAT instance creation has to be modified slightly, in particular for the modeling of the fault site. This is explained in the following.

The complete CNF $\Phi_{\mathscr{C}}$ of a circuit $\mathscr{C}$ is held during the SAT-based ATPG run using DCA. Clauses from $\Phi_{\mathscr{C}}$ are dynamically activated and deactivated. However, the circuitry has to be changed locally when targeting stuck-at faults as described in Sect. 4.1.1, since the faulty part has to be modeled, too. Extracting the entire faulty part for each fault is disadvantageous when using DCA, because this means overhead.

Therefore, two copies of the circuit are used: one copy for the good circuit $\mathscr{C}_G$ and one copy for the faulty circuit $\mathscr{C}_F$. Additionally, the D-chain constraints Φ_d for each line are included. More formally, the CNF for the circuit is composed as follows: $\Phi_{\mathscr{C}} = \Phi_{\mathscr{C}_G} \cdot \Phi_{\mathscr{C}_F} \cdot \Phi_d$. The modeling is shown in Fig. 5.11. By this, the DCA methodology activates clauses from the good and the faulty part of the circuit. The fault specific constraints Φ_F contain the CNF for the faulty gate Φ_f as well as the initial D-chain constraint Φ_f^d: $\Phi_F = \Phi_f \cdot \Phi_f^d$.

However, the fault site is included twice: one time as a non-faulty gate in $\Phi_{\mathscr{C}_F}$ and one time as a faulty gate in Φ_F. In order to avoid an inconsistent SAT instance, it has to be ensured that the non-faulty gate is not activated. This is done by flagging the corresponding variables as *inactive*. As a result, the clauses for this gate in $\mathscr{C}_F$ and its transitive fanin cone cannot be activated. The correct gate in $\mathscr{C}_F$ is substituted by the faulty gate. Furthermore, the inputs of the output cone have to correlate with the corresponding variables in the good circuit (see Fig. 4.2). Here, buffer constraints Φ_{buf} are inserted to guarantee that the inputs of the output cone have the correct value as depicted in Fig. 5.11. Additionally, the inputs of the output cone are also flagged as inactive. Otherwise, they have to be justified causing significant computational overhead. The following example demonstrates the SAT formulation.

Example 5.6. Consider the circuit shown in Fig. 5.12 (replicated from Fig. 4.3) Fig. 5.12a shows the good circuit and Fig. 5.12b shows the faulty circuit. Both circuits are identical at the beginning but have different variables. D-chain constraints have to be extracted for each line:

$$\Phi_{\mathscr{C}_G} = \Phi_a \cdot \Phi_b \cdot \Phi_o$$

$$\Phi_{\mathscr{C}_F} = \Phi_{a'} \cdot \Phi_{b'} \cdot \Phi_{o'}$$

$$\Phi_d = \Phi_{i_1}^d \cdot \Phi_{i_2}^d \cdot \Phi_{i_3}^d \cdot \Phi_{i_4}^d \cdot \Phi_a^d \cdot \Phi_b^d \cdot \Phi_o^d$$

Assume that there is a stuck-at fault $(i_1, 1)$. In order to generate a test pattern for $(i_1, 1)$, the fault-specific constraints are as follows:

$$\Phi_F = \Phi_a^{\text{faulty}} \cdot (\bar{i}_1) \cdot (i_1') \cdot (a^d)$$

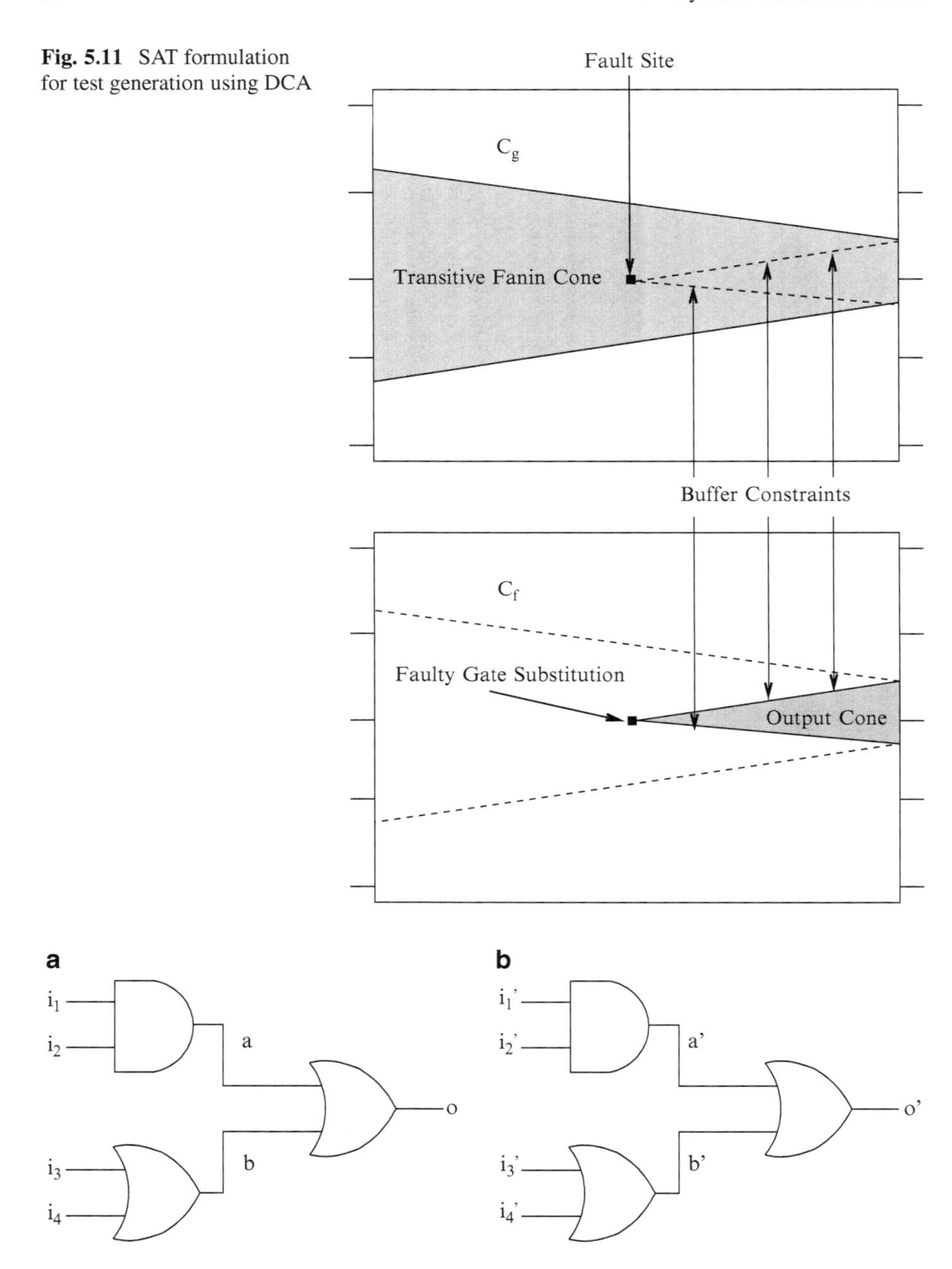

Fig. 5.11 SAT formulation for test generation using DCA

Fig. 5.12 Example circuit for DCA formulation. (**a**) Good circuit $\mathscr{C}_G$ and (**b**) faulty circuit $\mathscr{C}_F$

Variable i_1 has to be set to 0 and i'_1 has to be set to 1 to activate the fault. The d-variable a^d has to be 1 to begin fault propagation. The CNF Φ_a^{faulty} denotes the CNF for the faulty gate a[4]:

$$\Phi_a^{\text{faulty}} = (i'_1 + \overline{a}') \cdot (i_2 + \overline{a}') \cdot (\overline{i}'_1 + \overline{i}_2 + a')$$

Since line b/b' is an input to the output cone of fault $(i_1, 1)$, the line has to assume the same value in the good circuit as well as in the faulty circuit:

$$\Phi_{\text{buf}} = (b + \overline{b}') \cdot (\overline{b} + b')$$

Additionally, variable b' (input to output cone) as well as the variables i'_1 and a' (faulty gate) have to be flagged as inactive. This avoids activating the clauses for justifying the value in $\mathscr{C}_F$.

5.6.1 Dynamic Clause Activation and Multiple-Valued Logic

The section above presented the SAT formulation for Boolean circuits using DCA. However, as described in Chap. 4, Boolean logic is not sufficient when targeting industrial circuits. The technique has to be adjusted slightly in order to apply it to industrial circuits where a four-valued logic or hybrid logic, respectively, is used. This section deals with the DCA modeling of gates in four-valued logic.

In the DCA modeling for Boolean circuits, each clause ω_g of a gate g is registered in the SWL of the literal g or $\overline{g}$ – depending on the polarity of the literal contained in ω_g. The following problems are faced when using multiple-valued logic:

- Each line can be assigned not only one variable but multiple ones (representing the signal's value) due to the Boolean encoding.
- In the SAT formulation used for Boolean circuits, all clauses contain exactly one literal associated with the gate's output. This is different for industrial circuits. Here, more than one literal or no literal in a clause can be associated with the gate's output.

The SWL entries have to be adjusted in order to guarantee a consistent activation of the SAT instance. The following three cases must be considered for each clause ω:

- Clause ω contains exactly one literal associated with the gate's output. Then, nothing has to be changed and ω is registered as introduced in Sect. 5.3.
- Clause ω contains multiple literals $\lambda_1, \ldots, \lambda_n$ associated with the gate's output. Then, ω has to be registered in the entry of each literal $\overline{\lambda_1}, \ldots, \overline{\lambda_n}$ to make sure that the clause is activated and pushed on the J-stack if any assignment

[4]Note that all other inputs than the faulty line are taken from the correct circuit.

Table 5.3 CNF and SWL entries for a BUS gate using $\eta_{\mathscr{L}_4}$

Id	Entry	Clause	Id	Entry	Clause
ω_1	$c, \overline{c}^*$	$(a + a^* + \overline{c} + c^*)$	ω_8	$\overline{c}^*$	$(\overline{b} + \overline{b}^* + c^*)$
ω_2	$\overline{c}, c^*$	$(b^* + c + \overline{c}^*)$	ω_9	$\overline{c}^*$	$(\overline{a} + \overline{a}^* + c^*)$
ω_3	c	$(a + b + \overline{c})$	ω_{10}	$\overline{c}^*$	$(\overline{a}^* + \overline{b}^* + c^*)$
ω_4	$c, \overline{c}^*$	$(b + b^* + \overline{c} + c^*)$	ω_{11}	$\overline{c}$	$(\overline{b} + c)$
ω_5	c^*	$(a^* + b + \overline{b}^* + \overline{c}^*)$	ω_{12}	$\overline{c}$	$(\overline{a} + c)$
ω_6	c^*	$(a + \overline{a}^* + b^* + \overline{c}^*)$	ω_{13}	$\overline{a}, \overline{a}^*, b, \overline{b}^*$	$(a + a^* + \overline{b} + b^*)$
ω_7	c^*	$(\overline{a} + a^* + \overline{b} + b^* + \overline{c}^*)$	ω_{14}	$a, \overline{a}^*, \overline{b}, \overline{b}^*$	$(\overline{a} + a^* + b + b^*)$

is made to variables of the gate's output. In order to avoid that ω is added to Φ_{dyn} multiple times, a check has to be performed whether this clause is already activated. This induces some overhead on the activation procedure. However, since the proportion of these clauses is very small, the overhead is acceptable.

- Clause ω contains no literal associated with the gate's output but associated with the gate's inputs only. Then, ω has to be registered at each literal.

The following example shows how the clauses of a gate which is modeled in four-valued logic are registered in the SWL.

Example 5.7. Table 5.3 shows the CNF for a BUS gate with inputs a, b and output c. The clauses $\omega_1, \ldots, \omega_{12}$ represent the functionality of the BUS, whereas the clauses ω_{13} and ω_{14} prevent a bus conflict (see [vdL96]). Column *Entry* shows at which literal the clause is registered for activation. In contrast to the pure Boolean modeling, some clauses have to be registered by more than one literal, i.e. $\omega_1, \omega_2, \omega_4, \omega_{13}, \omega_{14}$. The clauses ω_{13} and ω_{14} prevent bus conflicts on the inputs. The output is not used in these clauses. Therefore, both clauses are registered at each of their literals.

Here, the flexibility of the DCA technique is clearly shown. Besides Boolean logic, it is easily possible to extend the technique to the more complex multiple-valued logic.

5.7 Experimental Results

This section presents the experimental results of the DCA technique. The SAT techniques proposed in this chapter were implemented in C++ on top of the SAT-based ATPG approach PASSAT as introduced in [DEF⁺08] resulting in the tool *DynamicSAT*. The SAT solver used in this approach is MiniSat [ES04] in version 1.14. Tests with the newer version, MiniSat 2, were also executed. However, this consistently resulted in worse run time. Therefore, MiniSat 1.14 was kept as the SAT engine. Furthermore, the publicly available circuit-based SAT solver CSAT [LWCH03, LWC⁺03] was also experimentally evaluated (see Sect. 3.5). However, the CNF-based SAT solver MiniSat turned out to be faster in solving ATPG problems.

Experimental results of DynamicSAT and competitive approaches are presented in this section. The largest circuits from the benchmarks suites ISCAS'85 [BF85], ISCAS'89 [BBK89] and ITC'99 [CSS00] are taken as benchmarks. Test generation is performed for two different fault models. Results for the SAFM – conducted on an AMD 64-bit Opteron (2.8 GHz, 32 GB, GNU/Linux) – are given in Sect. 5.7.1 and results for the PDFM – conducted on an AMD 64-bit Opteron (3.0 GHz, 64 GB, GNU/Linux) – are shown in Sect. 5.7.2. Since the techniques proposed in this book were developed in particular for industrial circuits, Sect. 5.7.3 introduces the industrial circuits in detail and gives experimental results.

5.7.1 Results for the Stuck-at Fault Model

DynamicSAT is compared to the structural ATPG approach Atalanta [LH93] as well as to the SAT-based ATPG approach TEGUS [SBS96] contained in SIS [SSL+92]. Furthermore, DynamicSAT is compared to the SAT-based ATPG tool PASSAT [DEF+08]. The SAT formulation of PASSAT resembles the SAT formulation of TEGUS, since only Boolean circuits are considered. However, PASSAT uses the modern SAT solver MiniSat to solve the CNFs. Note that TEGUS integrates a circuit-based learning feature for global implications which was disabled for these experiments. Results for TEGUS with learning enabled are given in the next chapter which focuses on circuit-based learning.

Test generation for the SAFM is performed without fault dropping. That is, no fault simulator is used to detect additional faults of a generated test pattern. Otherwise, the run times would have been too short for a practical comparison for the benchmark circuits. The abort criterion for the SAT-based approaches for each fault was 7 MiniSat restarts. A fault is marked as aborted when it could not be classified as either testable or untestable within this interval.

The results for each competitive approach are given in the respective column of Table 5.4. Atalanta is started with two different configurations. Column *Atalanta* reports the results for the default configuration of the tool, i.e. a backtrack limit of 10, while column *Atal. inc* gives the results for Atalanta with an increased backtrack limit of 100.[5] Columns entitled *Ab.* contain the number of faults which could not be classified in the given time interval. The CPU time for test generation is given in columns *Time* – either in CPU minutes (*min*) or CPU hours (*h*). The SAT-based approaches whose results are given in columns *TEGUS* and *PASSAT* have an additional column *%bld* which gives the proportion of the total run time needed for building the SAT instances.

First, the results of the structural ATPG algorithm and the classical SAT-based ATPG algorithms are discussed to show the shortcomings of both approaches. Atalanta is very fast especially for the small circuits. However, many faults

[5] The tool was not applicable for six benchmarks.

Table 5.4 Experimental results for the SAFM

	Atalanta		Atal. inc		TEGUS			PASSAT		
Circ.	Ab.	Time	Ab.	Time	%bld	Ab.	Time	%bld	Ab.	Time
c1355	42	0:01 min	0	0:01 min	73	0	0:03 min	86	0	0:02 min
c1908	8	0:01 min	3	0:01 min	74	0	0:02 min	88	0	0:03 min
c2670	77	0:01 min	70	0:01 min	79	0	0:03 min	74	0	0:05 min
c3540	3	0:01 min	1	0:01 min	77	0	0:09 min	88	0	0:11 min
c5315	6	0:02 min	0	0:02 min	80	9	0:11 min	84	0	0:08 min
c6288	1,450	0:09 min	1,143	0:20 min	64	55	1:48 min	51	0	1:21 min
c7552	269	0:08 min	211	0:12 min	80	28	0:22 min	84	0	0:15 min
s5378	0	0:02 min	0	0:02 min	84	0	0:04 min	91	0	0:02 min
s9234	53	0:10 min	48	0:11 min	85	80	0:19 min	89	0	0:10 min
s13207	12	0:25 min	9	0:25 min	93	6	0:32 min	90	0	0:11 min
s15850	23	0:29 min	6	0:41 min	88	3	1:00 min	87	0	0:28 min
s35932	*not appl.*		*not appl.*		99	0	2:56 min	61	0	0:14 min
s38417	12	5:24 min	4	5:26 min	96	0	3:50 min	89	0	0:32 min
s38584	*not appl.*		*not appl.*		98	0	3:40 min	90	0	0:18 min
b04	33	0:01 min	12	0:01 min	58	47	0:01 min	83	0	0:01 min
b11	0	0:01 min	0	0:01 min	73	0	0:01 min	90	0	0:01 min
b12	10	0:01 min	0	0:01 min	73	0	0:01 min	91	0	0:01 min
b13	0	0:01 min	0	0:01 min	75	0	0:01 min	93	0	0:01 min
b14	533	2:32 min	222	3:35 min	75	959	4:36 min	81	0	3:03 min
b15	2,715	2:47 min	1,778	6:33 min	63	1,728	7:39 min	87	0	4:32 min
b17	7,414	41:33 min	5,079	1:39 h	71	5,374	32:17 min	86	0	13:39 min
b18	*not appl.*		*not appl.*		86	7,992	3:21 h	75	0	1:03 h
b20	*not appl.*		*not appl.*		73	3,963	15:09 min	74	0	9:48 min
b21	*not appl.*		*not appl.*		73	3,276	15:52 min	74	0	10:15 min
b22	*not appl.*		*not appl.*		78	3,360	23:56 min	75	0	14:06 min
Av./total	12,660		8,586		79	26,871		82	0	

Table 5.5 Experimental results for the SAFM – DynamicSAT

	DynamicSAT					DynamicSAT Hybrid				
Circ.	%bld	%bits	Ab.	Time	Imp.	%bld	%bits	Ab.	Time	Imp.
c1355	11	85	0	0:01 min	2.00x	13	85	0	0:01 min	2.00x
c1908	28	47	0	0:01 min	3.00x	31	47	0	0:01 min	3.00x
c2670	26	13	0	0:01 min	5.00x	25	13	0	0:01 min	5.00x
c3540	39	29	0	0:01 min	11.00x	41	29	0	0:01 min	11.00x
c5315	42	17	0	0:01 min	8.00x	44	17	0	0:01 min	8.00x
c6288	4	84	2	1:34 min	0.86x	4	84	0	1:34 min	0.86x
c7552	27	18	0	0:03 min	5.00x	29	18	0	0:03 min	5.00x
s5378	33	29	0	0:01 min	2.00x	39	29	0	0:01 min	2.00x
s9234	40	22	0	0:02 min	5.00x	41	22	0	0:02 min	5.00x
s13207	31	13	0	0:03 min	3.67x	31	13	0	0:03 min	3.67x
s15850	57	9	0	0:05 min	5.60x	58	9	0	0:04 min	7.00x
s35932	22	31	0	0:08 min	1.75x	21	31	0	0:08 min	1.75x
s38417	30	21	0	0:12 min	2.67x	31	21	0	0:12 min	2.67x
s38584	29	24	0	0:10 min	1.80x	30	24	0	0:10 min	1.80x
b04	29	30	0	0:01 min	1.00x	30	30	0	0:01 min	1.00x
b11	54	40	0	0:01 min	1.00x	46	40	0	0:01 min	1.00x
b12	50	55	0	0:01 min	1.00x	56	55	0	0:01 min	1.00x
b13	50	40	0	0:01 min	1.00x	40	40	0	0:01 min	1.00x
b14	62	29	0	0:21 min	8.71x	62	29	0	0:21 min	8.71x
b15	12	15	128	2:55 min	1.55x	12	18	0	3:16 min	1.39x
b17	19	15	128	6:37 min	2.06x	15	16	0	8:42 min	1.57x
b18	3	25	11,566	3:07 h	0.34x	9	37	7	1:14 h	0.85x
b20	65	20	0	1:04 min	9.19x	65	20	0	1:04 min	9.19x
b21	67	21	0	1:04 min	9.61x	67	21	0	1:04 min	9.61x
b22	60	21	0	1:44 min	8.13x	60	21	0	1:44 min	8.13x
Av./total	36	30	11,824		4.04x	36	31	7		4.26x

cannot be classified when applied to larger circuits. The number of aborted faults decreases slightly at the cost of more run time when increasing the backtrack limit. TEGUS is much slower especially for the smaller circuits but produces less aborts. However, the number of aborted faults increases significantly for the large ITC'99 benchmarks. On average, 79% of the run time is spent for CNF generation. The percentage needed for CNF generation is even higher for PASSAT. Here, 82% of the run time is used for building the SAT instances. The increase can be explained with the improved SAT techniques. Less time is needed for the solving process. Due to the robustness of the modern SAT techniques, all faults can be classified, i.e. no aborts occur. PASSAT is generally much faster than TEGUS. Compared to Atalanta, PASSAT generally needs more run time for the small circuits due to the CNF generation, but performs very well on the larger benchmarks.

Table 5.5 presents the results for techniques proposed in this chapter. The results for DynamicSAT are shown in column *DynamicSAT*. Column *DynamicSAT Hybrid* gives the results for the hybrid approach emulating a classical SAT solver as described in Sect. 5.5. The average number of specified bits of the generated test is given in column *%bits*. This number is given in percentage of the average number of specified bits of tests generated by PASSAT. A low number of specified bits is particular important for test compaction. Column *Imp.* shows the run time improvement of DynamicSAT compared to PASSAT. Although holding the complete circuit CNF in memory, the memory consumption of DynamicSAT is negligible for these benchmarks and only slightly increased compared to PASSAT.

The percentage of time needed for SAT instance generation is much smaller compared to PASSAT. SAT instance generation using DynamicSAT consists mainly of the following parts: generation of the fault-specific constraints Φ_F, deactivating Φ_{dyn} and resetting the data structures of the SAT solver used. Activating the clauses is *not* part of the build time, but of the solving time. Only 36% on average is spent for SAT instance generation instead of 82%. It is also shown that large parts of the circuit are considered as IODCs because only 30% of the input bits are specified on average compared to PASSAT.

Furthermore, DynamicSAT reduces the run time of SAT-based ATPG significantly by an average factor of over 4. As drawback, the number of aborts increases significantly for a few more complex circuits. This is a result of the renunciation of the conflict-driven techniques. However, the aborts can nearly be eliminated applying the hybrid approach. Furthermore, the hybrid approach still improves the run time on average.

5.7.2 Results for the Path Delay Fault Model

Table 5.6 shows the experimental results for the PDFM. One million paths were chosen randomly for broadside test generation under the non-robust sensitization model. The DynamicSAT approaches are compared to the circuit-SAT-based approach KF-ATPG[6] [YCW04] and a modified version of PASSAT using the incremental features of MiniSat [ES04]. That is, all subsequent faults with the same output can share the SAT instance – as proposed in [CG96] – as well as the learned conflict clauses. Since all faults could be classified by all tools, no number of unclassified faults is given in the table.

The results show that KF-ATPG is much faster than the classical SAT-based ATPG approach PASSAT in most cases. This is due to the large percentage of build time needed by PASSAT. In spite of the incremental features used, 95% of the overall run time is spent on average for generating the SAT instances. Using DynamicSAT, the percentage of the build time decreases to only 58% of the overall run time which still seems to be high. However, considering the small total run time, this indicates that DynamicSAT is able to classify most faults very quickly.

[6] KF-ATPG was not applicable for one benchmark.

Table 5.6 Experimental results for the PDFM

	KF	PASSAT		DynamicSAT				DynamicSAT Hybrid			
Circ.	Time	%bld	Time	%bld	%bits	Time	Imp.	%bld	%bits	Time	Imp.
c1355	3:56 min	84	11:34 min	48	50	1:00 min	11.34x	48	50	1:03 min	11.02x
c1908	2:24 min	95	11:57 min	50	48	3:55 min	3.05x	50	48	3:51 min	3.10x
c2670	3:01 min	97	17:54 min	56	16	1:32 min	11.67x	56	16	1:30 min	11.93x
c3540	2:28 min	98	11:49 min	63	28	0:29 min	24.45x	63	28	0:28 min	25.32x
c5315	4:45 min	98	16:16 min	66	21	0:29 min	33.66x	66	21	0:29 min	33.66x
c6288	1:28 min	98	3:33 min	67	50	0:08 min	26.63x	65	50	0:07 min	30.43x
c7552	8:03 min	96	14:54 min	54	21	1:23 min	10.77x	54	21	1:22 min	10.90x
s5378	0:19 min	90	0:17 min	56	37	0:02 min	8.50x	53	37	0:02 min	8.50x
s9234	2:26 min	97	12:33 min	55	44	0:31 min	24.29x	56	44	0:30 min	25.10x
s13207	8:58 min	95	1:05 h	61	13	3:30 min	18.57x	62	13	3:27 min	18.84x
s15850	11:47 min	95	52:37 min	61	9	2:34 min	20.50x	62	9	2:32 min	20.77x
s35932	5:54 min	98	1:09 min	66	41	0:08 min	8.63x	65	41	0:08 min	8.63x
s38417	42:53 min	92	31:43 min	46	51	9:49 min	3.23x	46	51	9:54 min	3.20x
s38584	16:16 min	95	54:42 min	61	23	3:49 min	14.33x	61	23	3:46 min	14.52x
b04	0:40 min	96	1:45 min	48	50	0:08 min	13.13x	49	50	0:08 min	13.13x
b11	0:04 min	97	0:14 min	55	54	0:01 min	14.00x	54	54	0:01 min	14.00x
b12	*not appl.*	94	0:13 min	51	60	0:03 min	4.33x	50	60	0:03 min	4.33x
b13	0:01 min	93	0:01 min	57	62	0:01 min	1.00x	59	62	0:01 min	1.00x
b14	29:46 min	96	1:10 h	63	61	1:52 min	37.50x	64	61	1:49 min	38.53x
b15	19:54 min	97	1:44 h	81	11	0:50 min	124.80x	82	11	0:49 min	127.35x
b17	18:05 min	96	1:07 h	52	24	3:46 min	17.79x	52	24	3:46 min	17.79x
b18	1:41 h	94	4:51 h	61	7	29:12 min	9.97x	61	7	29:35 min	9.95x
b20	57:38 min	95	1:52 h	63	27	3:52 min	28.97x	63	27	3:52 min	28.97x
b21	59:02 min	95	1:46 h	63	27	3:47 min	28.02x	63	27	3:46 min	28.14x
b22	3:59 h	83	2:19 h	51	32	34:40 min	4.01x	51	32	34:57 min	3.98x
Average		95		58	35		20.13x	58	35		20.52x

Likewise to the SAFM, DynamicSAT reduces the average number of specified bits of the generated tests. Only 35% of the input bits set by PASSAT are specified using the DynamicSAT approach. The use of the DCA techniques has a significant impact on the run time. Compared to PASSAT, the test generation process can be accelerated by up to a factor of 124 and by a factor of over 20 on average (reported in Column *Imp.*).

The average run time improvement compared to KF-ATPG is still 8.67, although KF-ATPG learns unsatisfiable path segments and stores them in a circuit graph. Contrary to the SAFM, the application of the hybrid approach improves the run time only slightly, because most of the faults can be classified very quickly using DynamicSAT.

Table 5.7 Statistical information about industrial circuits – input/output

Circuit	PI	PO	FF	IO	Fixed
p44k	739	56	2,175	0	562
p49k	303	71	334	0	1
p57k	8	19	2,291	0	8
p77k	171	507	2,977	0	2
p80k	152	75	3,878	0	2
p88k	331	183	4,309	72	200
p99k	167	82	5,747	0	102
p177k	768	1	10,507	0	769
p456k	1,651	0	14,900	72	1,655
p462k	1,815	1,193	29,205	0	1,458
p565k	964	169	32,409	32	861
p1330k	584	90	104,630	33	519
p2787k	45,741	1,729	58,835	274	45,758
p3327k	3,819	0	148,184	274	4,438
p3852k	5,749	0	173,738	303	5,765

5.7.3 Results for Industrial Circuits

This section gives the experimental results for industrial circuits. First, some general information about the integration into an industrial framework are given. Then, the industrial circuits on which the experiments are conducted are introduced in detail. Finally, the results concerning the number of specified bits, run time and unclassified faults are given.

5.7.3.1 Integration into Industrial Framework

The resulting SAT-based ATPG approach was integrated into the industrial test environment of NXP Semiconductors as a prototype. All experiments were performed on an AMD 64-bit Opteron (2.8 GHz, 32 GB, GNU/Linux) using the static compaction flow with the SAT-based ATPG algorithm as test generator. Fault dropping is activated during test generation if not noted otherwise. When a test pattern is generated, the fault simulation detects all other faults that can be detected by this pattern. All of these detected faults are dropped from the fault list. No expensive test generation is required for the dropped faults. The abort criterion for each fault is 7 MiniSat restarts.

5.7.3.2 Industrial Circuits

Tables 5.7 and 5.8 show statistical information about the industrial circuits provided by NXP Semiconductors, Hamburg, Germany. All industrial circuits have been found to be difficult cases for test pattern generation.

Table 5.8 Statistical information about industrial circuits – size

Circuit	Gates	Fanouts	Tri	S-targets	T-targets
p44k	41,625	6,763	0	64,105	109,806
p49k	48,592	14,323	0	142,461	255,326
p57k	53,463	9,593	13	96,166	168,478
p77k	74,243	18,462	0	163,310	282,728
p80k	76,837	26,337	0	197,834	311,416
p88k	83,610	14,560	412	147,048	256,050
p99k	90,712	19,057	0	162,019	274,376
p177k	140,516	25,394	560	268,176	410,240
p456k	373,525	94,669	6,261	740,660	1,177,244
p462k	417,974	75,163	597	673,465	1,134,924
p565k	530,942	138,885	17,262	1,025,273	1,524,044
p1330k	950,783	178,289	189	1,510,574	2,464,440
p2787k	2,074,410	303,859	3,675	2,394,352	4,063,500
p3327k	2,540,166	611,797	44,654	4,557,826	6,620,254
p3852k	2,958,676	746,917	22,044	5,507,631	8,044,040

Table 5.7 gives input/output information. The name of the industrial circuit is given in the first column. Note that the name of the circuit roughly denotes the size of the circuit, e.g. p3852k contains over 3.8 million elements. Column *PI* presents the number of primary inputs of the circuit. The number of primary outputs is given in column *PO*. Column *FF* shows the number of state elements, i.e. flip-flops, while column *IO* denotes the number of input/output elements. Column *Fixed* gives the number of (P)PIs which are restricted to a fixed value. These inputs are not fully controllable.

Table 5.8 shows further information about the circuits' internals. Column *Gates* gives the number of gates in the circuit, while column *Fanouts* presents the number of fanout elements. Further, the number of tri-state elements, e.g. bus or busdrivers, is given in column *Tri*. Column *S-targets* (*T-targets*) shows the number of stuck-at (transition) fault targets after fault collapsing, i.e. the number of stuck-at (transition) faults that have to be tested.

Next, the experimental results of DynamicSAT for industrial circuits are presented. All experiments were conducted in the industrial ATPG framework of NXP Semiconductors with fault dropping enabled (contrary to the results for the benchmark circuits presented above). First, results concerning the test pattern compactness, i.e. the proportion of specified bits, are discussed. Then, a run time comparison of an industrial ATPG approach and the SAT-based approaches is given.

5.7.3.3 Results – Specified Bits

First, Table 5.9 shows the effect of the DCA technique on the proportion of specified bits. These results are also compared against the proportion of specified bits of the post-processor (see Sect. 4.2.3). Column *PASSAT Post* shows the results of the post-

Table 5.9 Experimental results – specified bits

	PASSAT Post		DynamicSAT			DynamicSAT Hybrid			
Circ.	Calls	%bits	Calls	%bits	Post (%)	Calls	Act. (%)	%bits	Post (%)
b14	3,385	33	3,406	50	44	3,406	0.0	50	44
b15	3,115	23	3,454	27	21	3,448	3.9	31	21
b17	8,729	28	9,209	29	25	9,106	1.8	31	25
b18	16,981	39	18,029	42	37	18,220	3.8	45	37
p44k	16,884	9	20,091	4	4	20,091	0.0	4	4
p49k	16,366	36	24,335	4	3	24,335	95.0	4	3
p57k	6,309	26	6,688	18	15	6,638	0.6	19	15
p77k	2,978	100	2,978	63	19	2,978	0.0	63	19
p80k	8,558	33	14,316	16	12	14,228	0.1	16	12
p88k	14,065	28	14,690	25	23	14,677	0.1	25	23
p99k	6,797	33	7,040	18	16	7,034	2.4	19	16
p177k	14,185	4	15,052	13	2	14,982	0.1	13	2
p456k	26,134	30	29,739	26	22	27,085	12.1	30	23
p462k	83,699	17	84,726	16	14	84,755	1.3	16	14
p565k	29,259	46	29,212	45	45	29,110	0.2	45	45
p1330k	42,568	19	42,197	20	12	42,227	0.3	20	12
p2787k	458,524	22	488,836	10	7	488,897	0.4	11	7
p3327k	311,631	14	328,093	6	5	316,677	1.6	6	5
p3852k	332,368	37	325,152	17	10	317,740	3.6	17	10
Average		30.4		23.6	17.7		6.7	24.5	17.7

processor. Columns *DynamicSAT* and *DynamicSAT Hybrid* give the results for the corresponding approaches.[7]

The number of the targeted faults are given in column *Calls* and the percentage of specified bits are presented in column *%bits*. This number is given in relation to the proportion of specified bits of the approach without applying the post-processor (column *Classic – w/o post*) in Table 5.9. Column *Post* gives the proportion of specified bits of approaches using DCA in combination with the subsequent application of the post-processor. DynamicSAT Hybrid has also an additional column *Act.* which gives the percentage of ATPG calls where the second phase, i.e. the SAT solver emulation, is reached. Here, the complete SAT instance is activated after four restarts. All variables are assigned (100%) for these target faults. Therefore, the average percentage of specified bits is slightly higher for the hybrid approach, in particular for those circuits where the percentage of a complete activation is high, e.g. p456k.

Comparing the number of specified bits of PASSAT with the post-processor and DynamicSAT, it can be observed that DynamicSAT produces fewer specified bits in most cases. Note that there is a difference between the benchmark circuits and the industrial circuits. The post-processor typically produces better results for the academic benchmark circuits, while DynamicSAT has generally fewer specified bits

[7] Here, the term *Hybrid* corresponds to the hybrid engine introduced in Sect. 5.5.

Table 5.10 Experimental results – competitive approaches

	FAN		FAN long		PASSAT			
Circ.	Ab.	Time	Ab.	Time	Mem.	Ab.	Time	Imp.F
b14	4	0:28 min	4	0:40 min	36M	0	0:39 min	0.97x
b15	624	1:04 min	483	6:54 min	33M	0	0:57 min	1.12x
b17	1,240	2:49 min	941	16:11 min	58M	0	2:29 min	1.13x
b18	475	5:14 min	315	10:34 min	149M	0	7:09 min	0.73x
p44k	0	1:13 min	0	1:13 min	61M	0	1:13 h	0.02x
p49k	2,719	4:26 h	1,808	5:39 h	96M	6,611	7:10 h	0.62x
p57k	197	1:21 min	138	6:55 min	80M	2	2:54 min	0.47x
p77k	0	0:07 min	0	0:07 min	90M	0	0:08 min	0.88x
p80k	18	1:41 min	12	2:40 min	111M	0	2:05 min	0.81x
p88k	56	1:36 min	35	2:34 min	102M	0	1:53 min	0.85x
p99k	988	1:29 min	455	6:05 min	108M	2	1:05 min	1.40x
p177k	99	2:13 min	59	3:32 min	207M	0	1:28 h	0.03x
p456k	5,964	21:50 min	3,524	2:14 h	513M	316	24:03 min	0.91x
p462k	1,009	9:17 min	785	11:06 min	445M	72	40:14 min	0.23x
p565k	263	8:41 min	175	11:21 min	586M	0	8:09 min	1.07x
p1330k	412	12:38 min	282	13:33 min	993M	0	31:44 min	0.40x
p2787k	218,292	2:21 h	165,699	11:53 h	1.86G	8,612	13:20 h	0.18x
p3327k	21,256	5:47 h	15,990	11:22 h	2.83G	2,939	20:13 h	0.29x
p3852k	23,447	7:47 h	17,247	11:44 h	3.27G	2,985	10:40 h	0.73x
Total	277,063		207,952			21,539		

for the industrial circuits. DynamicSAT (Hybrid) with the subsequent post-processor is able to further reduce the number of specified bits. This approach produces the best results for all circuits (except for benchmark circuit b14).

5.7.3.4 Results – Run Time

Next, a comparison of an industrial ATPG approach and the proposed SAT-based ATPG approaches are given. Table 5.10 shows the results of the industrial classical ATPG approach as well as the results of the SAT-based ATPG approach PASSAT in order to expose the advantages and weaknesses of each approach. The industrial *FAN* approach is a structural ATPG algorithm based on FAN [FS83]. This approach is highly optimized and used in an industrial ATPG framework. The approach *FAN long* corresponds to the FAN approach with the difference that a significantly increased backtrack limit and an increased time interval is used to reduce the number of aborts. The number of unclassified faults are shown in column *Ab.* and the run time is given in column *Time*. Furthermore, the results of PASSAT using the SAT solver MiniSat as well as the techniques presented in Sect. 4.2 are given in the table. The memory needs of PASSAT are presented in column *Mem.* and the run time improvement of PASSAT compared to the FAN approach is given in column *Imp.F*.

Table 5.11 Experimental results – dynamic clause activation

	DynamicSAT				DynamicSAT Hybrid			
Circ.	Mem.	Ab.	Time	Imp.P	Mem.	Ab.	Time	Imp.P
b14	40M	0	0:13 min	3.00x	40M	0	0:13 min	3.00x
b15	39M	3	0:53 min	1.08x	40M	0	1:07 min	0.85x
b17	91M	1	1:44 min	1.43x	91M	0	2:20 min	1.06x
b18	267M	29	9:14 min	0.77x	267M	0	6:53 min	1.04x
p44k	91M	0	5:11 min	14.08x	91M	0	5:10 min	14.13x
p49k	141M	16,291	8:16 h	0.87x	145M	16,291	13:01 h	0.55x
p57k	115M	8	0:59 min	2.95x	115M	3	0:58 min	3.00x
p77k	154M	0	0:08 min	1.00x	154M	0	0:08 min	1.00x
p80k	184M	0	1:25 min	1.47x	185M	1	1:26 min	1.45x
p88k	176M	0	1:31 min	1.24x	177M	0	1:31 min	1.24x
p99k	186M	16	1:17 min	0.84x	186M	9	1:16 min	0.86x
p177k	323M	7	5:50 min	15.09x	326M	0	5:58 min	14.75x
p456k	887M	6,987	1:35 h	0.25x	888M	1,282	59:09 min	0.41x
p462k	826M	390	17:41 min	2.28x	836M	95	35:11 min	1.14x
p565k	1.20G	0	18:02 min	0.45x	1.20G	0	18:00 min	0.45x
p1330k	1.82G	57	23:19 min	1.36x	1.84G	38	59:03 min	0.54x
p2787k	4.69G	589	6:30 h	2.05x	4.73G	45	6:45 h	1.98x
p3327k	5.51G	20,252	17:34 h	1.15x	5.62G	2,435	18:11 h	1.11x
p3852k	6.23G	21,458	14:37 h	0.73x	6.33G	11,796	15:07 h	0.71x
Total		66,088				31,995		

The results show that the structural ATPG is very fast but produces many aborts. Increasing the resources (FAN long) only leads to a slight decrease in the number of unclassified faults. At the same time, the run times increases. In contrast, PASSAT produces only few aborts but needs much run time. The run time is unreasonably high in some cases, e.g. p44k and p177k, which is not acceptable.

Table 5.11 shows the results for DynamicSAT as well as for DynamicSAT Hybrid. The factor of improvement compared to PASSAT is given in column *Imp.P*. The memory needs of DynamicSAT are not negligible anymore when applied to large circuits. The memory consumption of DynamicSAT is roughly twice as much for the larger circuits compared to the memory needs of PASSAT because of holding the complete CNF for the circuit in memory. The implementation of an efficient memory management and the development of strategies to reduce the memory consumption, e.g. by circuit partitioning or interval-based clause generation (and deletion), is future work.

DynamicSAT produces very few aborts compared to FAN. Nonetheless, there are more unclassified faults than PASSAT produces. The run time behavior of SAT-based ATPG is significantly improved in return. The highest speed-up factors are 15.09 for p177k and 14.08 for p44k. This is especially important since these are the circuits where PASSAT performs very poor. However, there are some circuits for which DynamicSAT needs more run time than PASSAT. DynamicSAT Hybrid is slightly slower but more robust than DynamicSAT. The hybrid approach reduces the number of unclassified faults by roughly a factor of 2. As a result, the total number of unclassified faults converges to the number of unclassified faults of PASSAT. The number of aborts is even significantly reduced in a few cases (p2787k in particular).

The run time of DynamicSAT and DynamicSAT Hybrid is in most cases only slightly improved by a factor between 1–3 or in some cases comparable or even worse than the run time of PASSAT. However, PASSAT needs high run times for some circuits as described above. This so far prevents the stand-alone application of SAT-based ATPG in industrial practice. The clear advantage of using DCA is that for those circuits where PASSAT needs unacceptably high run time, the run time can be reduced by more than an order of magnitude. As a result, the approaches using DCA have acceptable run time or perform very well, respectively, on *all* evaluated industrial circuits producing few aborts only.

In summary, the use of the DCA technique accelerates the SAT-based algorithm and is able to diminish or even close the run time gap between structural ATPG and SAT-based ATPG algorithms. At the same time, the advantages of SAT-based algorithms, in particular the high level of robustness, can be retained. Furthermore, the use of the DCA technique results in a significantly reduced portion of specified bits which is very important for test compaction.

5.8 Summary

Classical structural ATPG algorithms are usually very fast, but have problems to cope with hard-to-detect faults which occur more and more frequently in today's complex designs. Test generation based on Boolean Satisfiability (SAT-based ATPG) has shown to be a robust complement to structural ATPG. However, algorithms for SAT-based ATPG suffer from the overhead for easy-to-detect faults. Disadvantages of SAT-based ATPG are for instance the costly SAT instance generation, the missing support of ODCs and the loss of structural knowledge.

The SAT technique *Dynamic Clause Activation* (DCA) which addresses these drawbacks has been presented in this chapter. Using this technique, the CNF of a circuit is built only once and the needed parts of the CNF are dynamically activated during the solving process. Structural knowledge is used for a J-frontier as well as to model ODCs implicitly. The resulting tool *DynamicSAT* is able to reduce the run time of the ATPG process significantly. However, the robustness is slightly decreased. This is solved by *DynamicSAT Hybrid*. Here, a classical SAT solver is emulated to leverage the advantages of the conflict-driven SAT techniques due to the flexibility of DCA.

The experiments have shown that SAT-based ATPG is very robust for industrial circuits. However, classical SAT-based ATPG still needs unreasonably high run times for some circuits compared to structural ATPG which is not acceptable. This shortcoming has been shown to be overcome by the application of the DCA technique. The DCA technique is able to speed up SAT-based test generation significantly even for large circuits. This applies in particular to those circuits where the run time of classical SAT-based ATPG is high compared to structural ATPG. The SAT-based ATPG engine proposed in this chapter is fast and at the same time able to retain the high level of robustness of SAT-based ATPG. Furthermore, the portion of specified bits is very low and can further be improved by the combination with the post-processor.

Chapter 6
Circuit-Based Dynamic Learning

The proportion of unclassified faults produced by today's test generation algorithms grows due to the increased complexity of modern designs. However, a small percentage of unclassified faults is very important for the production test to keep a high fault coverage. Otherwise, faults may remain untested and defective devices could pass the test. A high fault coverage is needed to maintain a certain level of quality.

The application of SAT-based ATPG algorithms typically results in a significantly reduced number of unclassified faults compared to structural ATPG algorithms. SAT-based ATPG is very robust for hard-to-test faults due to the inherent learning strategies of modern SAT solvers. Although, many hard-to-test faults still remain unclassified in complex designs. A promising concept to strengthen the robustness of ATPG is the reuse of learned information.

This chapter presents efficient SAT techniques for reusing information [ED09]. Due to a tight integration into a SAT-based ATPG algorithm, dynamically learned conflict clauses can be efficiently passed from one target fault to another. By this, the number of unclassified faults can be significantly reduced in complex designs. Furthermore, the technique presented in this chapter can be easily combined with *Dynamic Clause Activation* (DCA) given in Chap. 5.

Keeping all learned information causes a large memory overhead and signifies an overhead for clause processing. Therefore, different strategies to identify unimportant information are presented to achieve a good trade-off between run time and the number of unclassified faults. Furthermore, the use of a *Post-Classification Phase* (PCP) is proposed. In this phase, faults are classified which were aborted particularly at the beginning of the ATPG run.

Section 6.1 describes the integration of the circuit-based dynamic learning technique into the SAT-based ATPG flow. Section 6.2 motivates and explains the use of the post-classification phase. Different learning strategies are presented in Sect. 6.3. Since the proposed dynamic learning techniques can be flexibly integrated into the DCA technique presented in the previous chapter, Sect. 6.4 shows an

S. Eggersglüß and R. Drechsler, *High Quality Test Pattern Generation and Boolean Satisfiability*, DOI 10.1007/978-1-4419-9976-4_6,

overview about the resulting SAT solving engine. Section 6.5 gives experimental results for benchmark circuits as well as for industrial circuits. The content of this chapter is summarized in Sect. 6.6.

6.1 Integration of Dynamic Learning

Classical ATPG algorithms typically use static learning concepts (see Sect. 2.4). Here, additional implications are identified on the circuit structure. This task is typically done as a pre-process. In contrast, state-of-the-art SAT solvers employ inherent dynamic learning as described in Sect. 3.3.2. They learn in form of conflict clauses. Conflict clauses are recorded dynamically during the search process. Every time a conflict occurs during the search, the conflict is analyzed and a conflict clause is generated, i.e. learning is performed. In a circuit-oriented problem, a conflict clause corresponds to a conflicting value assignment of connections. The recorded conflict clauses can be used by a SAT solver to derive additional implications efficiently. Much effort is spent for conflict clause generation and the search process benefits strongly from their use. *Pervasive* conflict clauses (or pervasive clauses in the following) are conflict clauses which depend on the circuit's function only and can be reused for each fault.

The efficient integration of the proposed circuit-based dynamic learning technique (or short: dynamic learning) in SAT-based ATPG algorithms is shown in this section. Contrary to the approach presented in [FWD07], an internal CNF-based database is used which avoids costly transformation steps from gate-level to CNF. Furthermore, the technique is independent from the fault ordering.

First, Sect. 6.1.1 introduces pervasive conflict clauses and shows their efficient identification. Then, Sect. 6.1.2 presents a watch-list strategy to store these clauses and a methodology to activate them. The combination with the DCA technique is presented in Sect. 6.1.3.

6.1.1 Pervasive Conflict Clause Identification

Conflict clause generation is done by resolution of those clauses which are responsible for the conflict (see Sect. 3.3.2). Such a clause avoids that the SAT solver reenters the non-solution search space again. By this, large parts of the search space are pruned and the search can be accelerated. As described in the previous chapter, the SAT instance Φ^F_{test} for test generation for fault F consists of two parts: the circuit part $\Phi_{\mathscr{C}}$ and the fault-specific constraints Φ_F. A conflict clause ω_C derived by resolution of the clauses $\omega_1, \dots, \omega_n$ is called pervasive if $\omega_1, \dots, \omega_n$ are contained in $\Phi_{\mathscr{C}}$ or they are themselves pervasive clauses. Pervasive clauses are said to be fault independent. These clauses can be reused in subsequent SAT instances to prune search space without running in the conflict first. On the other hand, conflict clauses which are generated using at least one clause out of Φ_F are *fault dependent*. Those have to be discarded after solving the SAT instance.

Crucial for an efficient transfer of pervasive clauses is a permanent variable assignment to connections. However, this causes some overhead in the search algorithm of the SAT solver which can be adjusted by a slight modification. Due to a permanent variable assignment, all pervasive clauses can be stored in the internal database in their original form. Thus, expensive transformation steps are avoided. Newly learned clauses can be stored efficiently inside the SAT solver by using pointers.

Now, the identification of pervasive clauses is described. When building the SAT instance for fault F, distinguishing between fault independent clauses $\Phi_{\mathscr{C}}$ and fault dependent clauses Φ_F is easily possible. All learned clauses that have their source in at least one fault dependent clause have to be discarded and cannot generally be reused for subsequent SAT instances. Two methods are used:

- *ID identification* – Each variable gets an ID (corresponding to the variable's name) when assigned to a connection. The IDs are allocated in increasing order. The variable with the highest ID is defined as a fixed upper limit u after the permanent variable assignment has been done. If new variables have to be defined for Φ_F, e.g. describing faulty circuitry, only variables with an ID larger than u are used. Thus, the learned clause is discarded if the clause contains at least one variable with an ID larger than u.
- *Literal identification* – The literal λ^F is added to each fault dependent clause ω that does not use only new (fault dependent) variables. The literal λ^F is set to false by an *incremental assumption* [ES04] before each run. By this, ω keeps the original meaning – λ^F is redundant. Because λ^F is only used in one phase, each conflict clause derived from ω contains also λ^F since λ^F cannot be eliminated by resolution. These clauses can easily be identified as fault dependent clauses.

The following example demonstrates the use of literal identification.

Example 6.1. Consider again Example 3.4. Here, a conflict clause is generated by the application of resolution. The clauses $\omega_1 = (a+\overline{b})$, $\omega_2 = (b+\overline{c}+d)$ and $\omega_3 = (\overline{a}+d+e)$ are used. Assume that ω_2 is fault-specific and part of Φ_F. Therefore, the literal λ^F is added to this clause:

$$\omega_2 = (b+\overline{c}+d+\lambda^F)$$

When the resolution operator is applied as done in Example 3.4, the following conflict clause results:

$$\omega_C = (\overline{c}+d+e+\lambda^F)$$

Since λ^F is contained in the derived clause, ω_C is identified as a fault dependent clause and, consequently, non-pervasive clause. Therefore, ω_C has to be discarded.

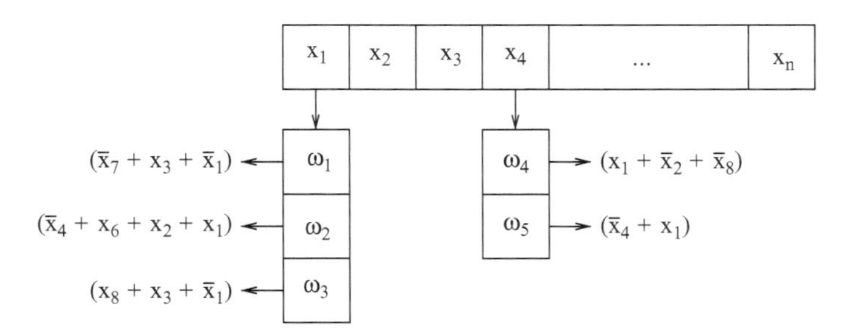

Fig. 6.1 Illustration of a watch list example

In summary, each learned clause $\omega = (\lambda_1 + \ldots + \lambda_n)$ is checked after each run and discarded if the following holds:

$$\bigvee_{i=1}^{n} ((\lambda_i = \lambda^F) \vee (\mathrm{ID}(\lambda_i) > u))$$

Otherwise, ω is a pervasive clause and can be reused for each subsequent fault.

6.1.2 Variable-Based Activation

Not all stored clauses should be reused in each SAT instance. Subsequent SAT instances may target faults from other regions of the circuit. A pervasive clause learned from one part of the circuit is useless if that part is not included in the current SAT instance. Due to the large number of pervasive clauses generated during ATPG for large designs, checking each single clause is not feasible and may outweigh the benefit. Therefore, the concept of *variable-based activation* is introduced. Here, the internal database is modeled as a watch list (similar to the watch list proposed in Sect. 5.2.2). A list of clauses is assigned to each variable x_i of the circuit.

These clauses are said to be "watched" by x_i. When a new pervasive clause

$$\omega_p = (\lambda_1 + \ldots + \lambda_n)$$

is learned, one variable x_{λ_i} contained in ω_p is chosen. Then, ω_p is added to the list of x_{λ_i}, i.e. ω_p is "watched" by x_{λ_i}. In the following, the set of clauses which is watched by variable x_{λ_i} is denoted by $\Phi(x_{\lambda_i})$. A variable-based watch list which is very similar to the watch list presented for DCA in Fig. 5.5 is illustrated in Fig. 6.1 as an example. The variables of the circuit are denoted by $x_1, \ldots, x_n$ and $\omega_1, \ldots, \omega_5$ are the pervasive clauses stored in the internal database.

By this, the set of pervasive clauses for Φ^F_{test} can be efficiently determined. All learned clauses which are "watched" by variables contained in Φ^F_{test} can be added directly to Φ^F_{test}. In other words, when the CNF Φ_g for gate g with output variable x_g is added to Φ^F_{test}, the "watched" clause set $\Phi(x_g)$ is added, too. The clauses of

$\Phi(x_g)$ are said to be *active* for Φ^F_{test}. Those clauses which come from other parts of the circuit do not have to be considered and, therefore, do not have to be processed. These clauses are *inactive* for Φ^F_{test}.

However, a learned pervasive clause ω_p is only useful for Φ^F_{test} if all variables are contained in Φ^F_{test}. Otherwise, at least one variable remains unconstrained and can satisfy ω_p. Nonetheless, "watching" a learned clause by one variable only is sufficient. This is because the overhead of determining whether all variables are contained in Φ^F_{test} is higher than adding some "useless" clauses to Φ^F_{test}.

6.1.3 Combination with Dynamic Clause Activation

Since the variable-based activation proposed for dynamic learning is very similar to the activation methodology proposed for DCA (see Chap. 5), circuit-based dynamic learning can be combined with DCA. The presented learning scheme can easily be integrated in the activation methodology of the DCA technique. For this purpose, an additional literal-based watch list has to be used besides the *Structural Watch List* (SWL) to keep track of the learned clauses. This watch list is named *Learned Watch List* (LWL), because the list only contains those learned pervasive clauses which should be dynamically activated during the search process. The LWL is located in the circuit part of the proposed search engine (see Sect. 5.1).

Variable-based activation is substituted by literal-based activation (as described in Sect. 5.3) in order to fit in the DCA framework. In contrast to the clauses from $\Phi_{\mathscr{C}}$, a pervasive clause ω_p does not have to be registered at a specific literal, i.e. the literal denoting the output of the gate. An arbitrary literal of ω_p can be chosen for registering the clause in the LWL, since a conflict clause is redundant and "only" prunes search space.

None of the learned clauses is activated before the search process starts. The learned clauses are dynamically added to Φ_{dyn} during the search process. When an activation request for a literal λ is sent, not only the clauses from the SWL are activated, but additionally those watched by λ in the LWL. Learned clauses registered at literals from other parts of the circuit are explicitly *not* touched. The same holds for clauses registered at $\overline{\lambda}$ (unless the assignment of the variable becomes negated), because these are implicitly satisfied. The activated learned clauses are treated differently only in one way. Because learned clauses are redundant, they are ignored in the break condition. That is, these clauses are not pushed on the J-stack $\mathscr{J}$ (see Sect. 5.4.1) and do not result in overhead for checking the break condition. The following example demonstrates the activation of learned clauses:

Example 6.2. Consider the circuit shown in Fig. 6.2. Assume that the pervasive conflict clause $\omega_p = (\overline{o}_1 + \overline{i}_2)$ was learned in a previous run. The conflict clause is registered in the LWL in the list of literal i_2. When the assignment $i_2 = 1$ occurs

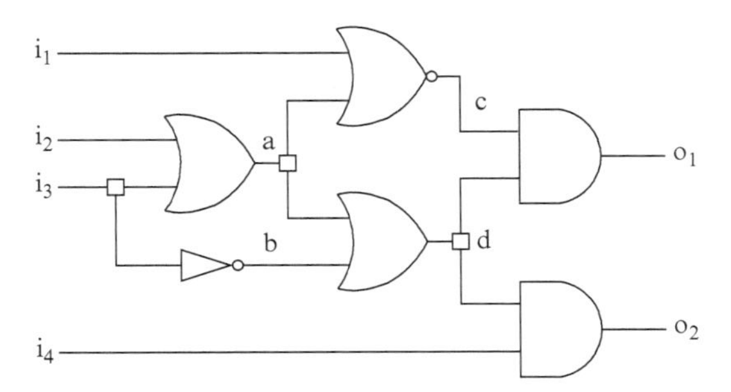

Fig. 6.2 Example circuit $\mathscr{C}$ for dynamic learning

during the search process, ω_p is activated and $o_1 = 0$ is directly implied. In case of the assignment $i_2 = 0$, ω_p would be satisfied and the search process does not benefit from the activation of ω_p. Therefore, ω_p remains deactivated.

However, the DCA technique has to be slightly modified when integrating the dynamic learning technique. A conflict clause is not bound to a specific region of the circuit, but denotes invalid signal correlations. When a conflict clause is activated which contains literals from parts of the circuit which do not have to be considered for the current target fault, other regions can be activated. To avoid this, all variables which are located in the circuit part relevant for the targeted fault are flagged. Flagging a variable means that clauses watched by this variable are allowed to be activated. Activation requests for non-flagged variables are ignored during the search process. The overhead for the flagging procedure is negligible because efficient caching strategies can be used.

6.2 Post-Classification Phase

The use of a post-classification phase is motivated and described in this section. Consider the diagram shown in Fig. 6.3. In this diagram, the progress of the number of learned clauses of ATPG for transition faults for the industrial design p57k is shown. On the x-axis, the progress in terms of targeted faults is denoted. The number of learned clauses is presented on the y-axis. The upper line shows the total number of learned clauses and the crosses present the number of "active" clauses for each target fault.

The diagram clearly shows that faults targeted at the beginning of the ATPG run have access to a smaller number of learned clauses than faults targeted at the end of the ATPG run. Furthermore, experimental evaluations have shown that – when using dynamic learning – most of the remaining unclassified faults appear at the beginning of the ATPG run. The ordering of the faults clearly has an influence on the classification. The effect can be mitigated by using a post-classification phase. The post-classification phase starts after all faults were targeted. In this phase, all faults that were aborted in the ATPG run are targeted again. The benefit is that – in

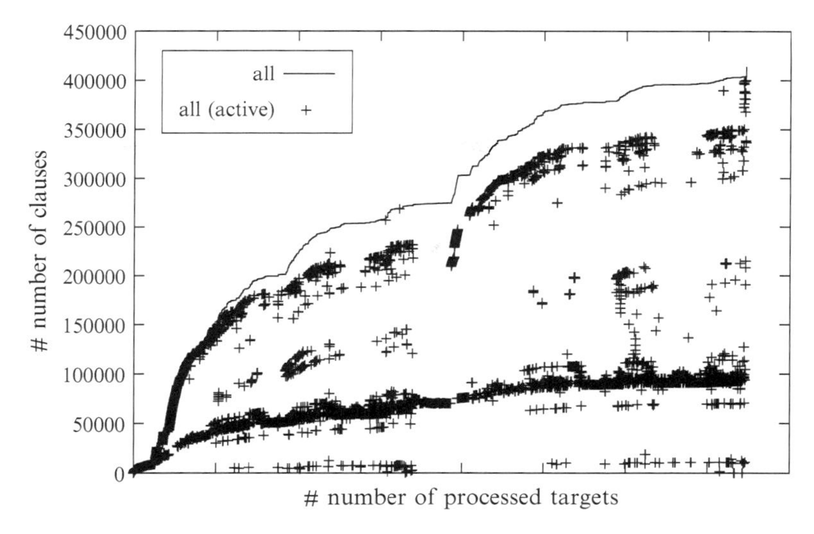

Fig. 6.3 Progress of (active) learned clauses – p57k

most cases – the search algorithm has access to significantly more learned pervasive clauses than in the previous ATPG call for these faults. Therefore, it is likely that many faults that were aborted in the previous phase can now be classified since the learned clauses already prune parts of the search space. Optionally, the time limit can be decreased to save run time.

6.3 Learning Strategies

SAT solvers typically generate a large number of conflict clauses during their search, especially when performing ATPG for large circuits. Keeping all learned pervasive clauses results in a large memory overhead. Furthermore, the efficiency of the BCP routine of the SAT solver can be decreased by too many learned clauses since all these clauses have to be processed. The clause size was proposed as a selection criterion in TG-GRASP [MS97]. Any clause that contains more literals than a user-defined maximum is discarded. The user-defined maximum of TG-GRASP was set to 10. In [FWD07], the maximum number was set to only three literals in a clause. This section introduces an additional selection criterion.

The problem of the excessive number of conflict clauses does not only exist in conjunction with the reuse of learned information but with conflict clause generation in general. Therefore, state-of-the-art SAT solvers have an inherent feature to judge the importance of a recorded conflict clause. The importance of a conflict clause is defined over the frequency of being involved in conflicts. Therefore, all recorded conflict clauses have an *activity value* [GN02].

This value is increased every time a conflict clause is used during conflict analysis. Conflict clauses with a high activity value are often involved in conflicts. Therefore, these clauses are considered important. In contrast, conflict clauses with a low activity value are less important. This measurement is used to reduce the size of the clause database. Clauses with a low activity are deleted in intervals. In order to take the search progress in account, each activity value is also reduced in specific intervals. By this, the importance of conflict clauses which had a high activity value in already traversed parts of the search space but are not useful anymore is decreased.

The activity measurement is proposed to serve as an additional selection criterion when using circuit-based dynamic learning. Therefore, the following strategies are chosen for experimental evaluation:

- *All* – No selection criterion is employed and all recorded pervasive conflict clauses are kept in the internal database.
- *Clause size* – All pervasive conflict clauses containing no more literals than a user-defined number are kept in the internal database. Here, reasonable clause sizes for experimental evaluation are 3,10 and 20.
- *Activity* – All pervasive clauses are ordered according to their activity after each run. Then, those pervasive clauses with the highest activity are reused and stored in the internal database. Reasonable clause set sizes for experimental evaluation are 1/4, 1/3 and 1/2 of all recorded clauses.

6.4 Improved SAT Solving Engine

The combination of the proposed circuit-based dynamic learning technique with the dynamic clause activation technique described in the previous chapter results in a novel ATPG-specific SAT solving engine. This section gives an overview on this engine which is illustrated in Fig. 6.4. At first, the complete circuit CNF is built and stored together with structural information which is used for dynamic clause activation. Then, the fault-specific constraints are extracted separately for each fault and the solving process is started using parts of the circuit CNF dynamically. After each run, the learned information is extracted according to the learning strategy presented above and stored in the internal database of the search engine using literal-based activation. The learned information can be dynamically activated for subsequent faults using the same activation methodology than for the circuit CNF.

As the experiments show, the proposed solving engine is able to accelerate SAT-based ATPG significantly that the run time gap between structural ATPG and SAT-based ATPG can be diminished or even closed as discussed in the previous chapter. In addition, the robustness of a SAT-based algorithm is retained in spite of the modifications and the number of unclassified faults can be reduced to a minimum due to the reuse of learned information.

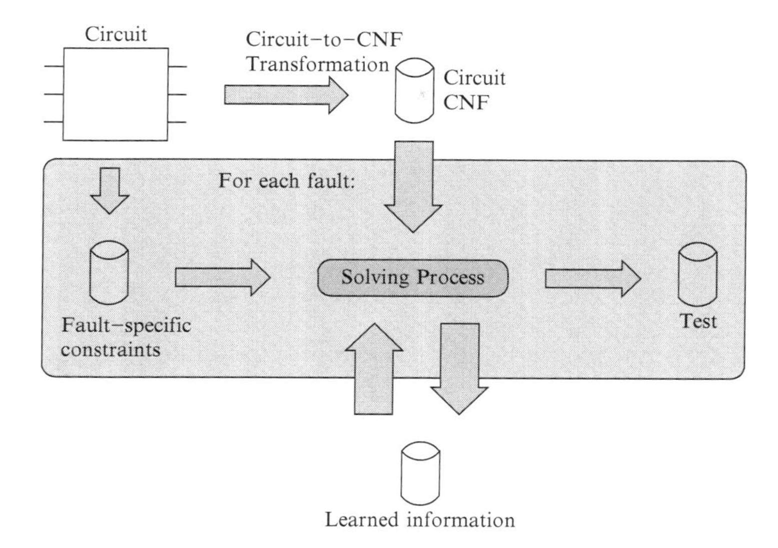

Fig. 6.4 Improved search engine

6.5 Experimental Results

The proposed circuit-based dynamic learning technique was integrated into the SAT-based ATPG framework such that the technique can be used by PASSAT as well as by DynamicSAT and DynamicSAT Hybrid. These approaches are extended with a '+' in the following, i.e. PASSAT+, DynamicSAT+ and DynamicSAT+ Hybrid to distinguish them from the approaches without dynamic learning. As described in Sect. 6.1.1, PASSAT had to be modified that the variable assignment to the connections is permanent for all targeted faults. This is implicitly given in DynamicSAT and DynamicSAT Hybrid. The experimental results of this technique are presented in this section.

The aim of the application of the dynamic learning technique is to decrease the number of unclassified faults. However, due to the large amount of additional information, i.e. conflict clauses, the solving process is typically slower since more clauses have to be processed during BCP. As shown in the previous chapter, SAT-based ATPG is robust enough to classify nearly all faults of the considered non-industrial benchmark circuits. As a result, the full advantage of the proposed technique cannot be shown on the benchmark circuits but only on the industrial circuits.

First, experimental results for the benchmark circuits are shown in Sect. 6.5.1 (SAFM) and Sect. 6.5.2 (PDFM). Section 6.5.3 gives the impact of this technique on industrial circuits. Furthermore, Sect. 6.5.4 shows that the resulting SAT-based ATPG engine can even improve the combination of structural ATPG and SAT-based ATPG as described in Sect. 4.3.

6.5.1 Dynamic Learning for Stuck-at Faults

The results for test generation for the SAFM for benchmark circuits are presented in this section. The experimental setup follows the setup used in Chap. 5. The results are presented in Table 6.1. Here, three different approaches with learning features are compared. The number of aborts of each approach is given in column *Ab.* Column *Time* gives the run time needed for all targeted faults. The improvement of the learning feature is presented in column *Imp.* with respect to the results of the corresponding approach without learning presented in Tables 5.4 and 5.5. *TEGUS+* [SBS96] learns global implications on the circuit structure. *PASSAT+* and *DynamicSAT+ Hybrid* correspond to the approaches introduced in the previous chapter with the proposed circuit-based dynamic learning feature integrated. Here, the learning strategy *all* is used. That is, all pervasive clauses are stored for reuse.

The learning feature of TEGUS is quite effective for the smaller benchmarks. Nearly all aborted faults can be classified using learning. The run time overhead is moderate for these benchmarks. However, the run time overhead for the larger benchmarks b14–b22 is not acceptable. Test generation for b15–b22 is aborted after 36 CPU hours. PASSAT+ with the integration of the presented dynamic learning technique has a slight increase in run time on average and can classify all circuits completely. However, PASSAT was already able to classify all faults without circuit-based learning. The few aborted faults of the hybrid DynamicSAT approach (see Table 5.5) can be eliminated using the proposed learning technique. However, the amount of learned information reduces the performance of the approach, in particular for the larger circuits.

As pointed out above, the advantage of the integration of circuit-based learning could not be shown on the smaller benchmark circuits. However, the test generation process of DynamicSAT Hybrid is analyzed for circuit b18 in order to show the effectiveness of the technique. The circuit b18 was chosen because it contains the most hard-to-test faults among the benchmark circuits. Table 6.2 presents the number of classified faults with respect to the needed restarts with and without dynamic learning. Each row shows how many faults can be classified within the corresponding number of restarts (column *Restarts*).

Column *Without learning* gives the number of classified faults without applying the proposed learning technique and column *With learning* gives information about how many faults could be classified within the interval with the learning technique enabled. The difference between both approaches is presented in column *Difference*. For example, 215,617 faults could be classified after one restart when learning is disabled. With dynamic learning enabled, the number grows to 234,992 faults in this restart slot. In contrast, the number decreases from 30,576 to 17,085 for restart slot 5.

The table clearly shows that generally less restarts, i.e. less conflicts and backtracks, are needed to classify a fault. Therefore, the search space can be traversed more quickly. Nonetheless, a general run time increase can be observed. This is

Table 6.1 Experimental results for circuit-based dynamic learning – stuck-at

	TEGUS+			PASSAT+			DynamicSAT+ Hybrid		
Circ.	Ab.	Time	Imp.	Ab.	Time	Imp.	Ab.	Time	Imp.
c1355	0	0:03 min	1.00x	0	0:02 min	1.00x	0	0:01 min	1.00x
c1908	0	0:02 min	1.00x	0	0:03 min	1.00x	0	0:01 min	1.00x
c2670	0	0:03 min	1.00x	0	0:05 min	1.00x	0	0:01 min	1.00x
c3540	0	0:09 min	1.00x	0	0:11 min	1.00x	0	0:01 min	1.00x
c5315	6	0:19 min	0.58x	0	0:08 min	1.00x	0	0:01 min	1.00x
c6288	0	2:00 min	0.90x	0	1:19 min	1.03x	0	5:43 min	0.27x
c7552	0	0:25 min	0.88x	0	0:16 min	0.94x	0	0:03 min	1.00x
s5378	0	0:04 min	1.00x	0	0:02 min	1.00x	0	0:01 min	1.00x
s9234	3	1:46 min	0.18x	0	0:10 min	1.00x	0	0:02 min	1.00x
s13207	0	0:32 min	1.00x	0	0:11 min	1.00x	0	0:04 min	0.75x
s15850	0	1:01 min	0.98x	0	0:29 min	0.97x	0	0:05 min	0.80x
s35932	0	2:55 min	1.01x	0	0:13 min	1.08x	0	0:09 min	0.89x
s38417	0	3:52 min	0.99x	0	0:32 min	1.00x	0	0:13 min	0.92x
s38584	0	3:41 min	1.00x	0	0:18 min	1.00x	0	0:12 min	0.83x
b04	0	0:03 min	0.33x	0	0:01 min	1.00x	0	0:01 min	1.00x
b11	0	0:01 min	1.00x	0	0:01 min	1.00x	0	0:01 min	1.00x
b12	0	0:01 min	1.00x	0	0:01 min	1.00x	0	0:01 min	1.00x
b13	0	0:01 min	1.00x	0	0:01 min	1.00x	0	0:01 min	1.00x
b14	1	1:08 h	0.07x	0	3:05 min	0.99x	0	0:34 min	0.62x
b15	*Timeout*		–	0	4:37 min	0.98x	0	12:58 min	0.25x
b17	*Timeout*		–	0	14:11 min	0.96x	0	23:33 min	0.37x
b18	*Timeout*		–	0	1:05 h	0.96x	0	2:26 h	0.51x
b20	*Timeout*		–	0	10:25 min	0.94x	0	1:53 min	0.57x
b21	*Timeout*		–	0	10:28 min	0.98x	0	1:36 min	0.67x
b22	*Timeout*		–	0	14:37 min	0.96x	0	3:02 min	0.57x

Table 6.2 Number of classified faults with respect to the number of needed restarts for b18

Restarts	Without learning	With learning	Difference
0	1,402	1,635	+233
1	215,617	234,992	+19,375
2	17,710	9,430	−8,280
3	8,319	8,917	+598
4	4,309	5,872	+1,563
5	30,576	17,085	−13,491
6	2	4	+2

caused by the overhead of the BCP procedure. BCP has to process significantly more clauses. Therefore, this observation motivates the use of learning strategies to reduce the total number of learned clauses and keep only the relevant ones.

The application of the circuit-based dynamic learning technique does not show a large impact on the benchmark circuits on the first sight. However, these circuits can already be classified completely using the existing methods. Furthermore, the amount of learned information reduces the overall performance of SAT-based ATPG. However, an analysis of the benchmark b18 shows that the search space can be traversed more quickly, i.e. the solution can be found with less backtracks.

Table 6.3 Experimental results for circuit-based dynamic learning – path delay

	PASSAT+		DynamicSAT+		DynSAT+ Hyb.	
Circ.	Time	Imp.	Time (m)	Imp.	Time (m)	Imp.
c1355	11:32 min	1.00x	1:17	0.78x	1:15	0.84x
c1908	11:55 min	1.00x	1:58	1.99x	1:58	1.98x
c2670	17:50 min	1.00x	1:00	1.53x	1:00	1.50x
c3540	11:50 min	1.00x	0:40	0.73x	0:40	0.70x
c5315	16:33 min	0.98x	0:35	0.83x	0:35	0.83x
c6288	3:34 min	1.00x	0:10	0.80x	0:10	0.70x
c7552	15:02 min	0.99x	0:59	1.41x	0:59	1.39x
s5378	0:17 min	1.00x	0:02	1.00x	0:02	1.00x
s9234	12:23 min	1.01x	0:29	1.03x	0:30	1.00x
s13207	1:06 h	0.98x	5:44	0.61x	5:46	0.60x
s15850	52:19 min	1.01x	2:34	1.00x	2:36	0.97x
s35932	1:13 min	0.95x	0:12	0.67x	0:12	0.67x
s38417	31:12 min	1.02x	7:29	1.31x	7:33	1.31x
s38584	55:29 min	0.99x	6:58	0.55x	7:08	0.53x
b04	1:45 min	1.00x	0:08	1.00x	0:08	1.00x
b11	0:14 min	1.00x	0:01	1.00x	0:01	1.00x
b12	0:13 min	1.00x	0:02	1.50x	0:02	1.50x
b13	0:01 min	1.00x	0:01	1.00x	0:01	1.00x
b14	1:09 h	1.01x	1:41	1.11x	1:41	1.08x
b15	1:43 h	1.01x	0:57	0.88x	0:57	0.86x
b17	1:07 h	1.00x	2:28	1.53x	2:27	1.54x
b18	5:04 h	0.96x	9:28	3.08x	8:52	3.34x
b20	1:49 h	1.03x	3:01	1.28x	3:05	1.25x
b21	1:46 h	1.00x	3:46	1.29x	3:47	1.28x
b22	2:45 h	0.84x	19:35	1.77x	19:22	1.80x

6.5.2 *Dynamic Learning for Path Delay Faults*

This section presents the results of the circuit-based dynamic learning techniques for the PDFM. Table 6.3 shows the results of three approaches integrating the learning technique. Since all faults could be classified completely (already without dynamic learning), the presentation of the results focuses on the performance. Here, a difference between PASSAT+ and DynamicSAT+ can be observed. The integration of the proposed dynamic learning technique has only marginal effects on the run time. Both DynamicSAT approaches, however, perform better with dynamic learning integrated. The run time is reduced in particular for the larger ITC'99 benchmarks. For instance, the performance on b18 is improved by a factor of 3.34.

This is contrary to the application of dynamic learning to stuck-at fault test generation. However, this can be explained with the different fault modeling. The fault-specific constraints for stuck-at test generation are larger than for path delay test generation. The fault-specific constraints for the PDFM include fixed value assignments only. Learning for stuck-at test generation is restricted to the good circuit part in this approach. In the faulty part, no learned information can be used. However, the ATPG algorithm uses only the good circuit part for path delay fault test generation, i.e. there is no faulty part. Therefore, the algorithm is able to use learned information in the complete SAT instance. As a result, more pervasive conflict clauses can be reused.

In summary, the use of the presented circuit-based dynamic learning techniques already results in slight run time improvements on average for these benchmark circuits.

6.5.3 Dynamic Learning for Industrial Circuits

This section shows the evaluation of the dynamic learning techniques on industrial circuits. Furthermore, results for the post-classification phase are presented. This was not possible before since the benchmark circuits used above could be already fully classified. The following learning strategies were evaluated next as proposed in Sect. 6.3:

- *All* – No selection criterion is employed and all recorded pervasive conflict clauses are kept in the internal database.
- *Clause size* – All pervasive conflict clauses containing not more literals than a user-defined number are kept in the internal database. Extensive experimental studies have been shown that the best performance could be achieved with a *Clause size* of 10. Therefore, only these results are given in detail.
- *Activity* – All pervasive clauses are ordered according to their activity after each run. Then, those pervasive clauses with the highest activity are reused and stored in the internal database. In this experimental evaluation, storing 1/2 of all recorded clauses for reuse turned out to provide best performance.

After presenting the results of the learning strategies, the results for the post-classification phase are given.

6.5.3.1 Learning Strategies

The experiments were carried out using PASSAT as well as DynamicSAT Hybrid which is more robust than DynamicSAT. All learning strategies presented above were experimentally evaluated. The results are summarized in Table 6.4. Detailed results for those strategies which turned out to be best are presented below. Table 6.4 also shows statistical data about the learned pervasive clauses as well as about the memory overhead and the total number of unclassified faults.

Column *Max* gives the maximum number of pervasive clauses learned during generation of a test for one fault. Column *Total* sums up the total number of clauses learned for each circuit and column *Average* shows the average number of clauses learned for each fault. The memory overhead is given in relation to the

Table 6.4 Summary of the experimental evaluation of all learning strategies

	PASSAT+					DynamicSAT+ Hybrid				
Strategy	Max	Total	Av.	Mem.	Ab.	Max	Total	Av.	Mem.	Ab.
None	–	–	–	–	21,539	–	–	–	–	31,995
All	1,001	313k	3.9	1.08x	987	1,189	843k	40.4	1.66x	728
Size 3	275	150k	0.6	1.00x	2,315	725	417k	7.7	1.03x	1,318
Size 10	549	173k	1.2	1.01x	1,270	892	512k	11.6	1.04x	1,009
Size 20	705	195k	1.6	1.01x	1,010	952	570k	14.3	1.06x	1,043
Act 1/4	368	215k	1.0	1.03x	1,102	668	396k	13.9	1.17x	1,401
Act 1/3	418	237k	2.4	1.05x	1,061	723	450k	17.4	1.25x	1,119
Act 1/2	584	246k	2.9	1.06x	961	798	604k	23.4	1.37x	1,991

memory consumption of the corresponding approaches without dynamic learning in column *Mem.* Column *Ab.* gives information about the total number of unclassified faults produced by this strategy.

The results show that PASSAT+ and DynamicSAT+ Hybrid behave differently with respect to learned information. The amount of learned clauses of DynamicSAT+ Hybrid is constantly larger than the amount of learned clauses of PASSAT+. This can be explained by the focusing on conflict-driven techniques in the used SAT solver MiniSat. The search is guided by conflict-driven measurements and the recorded conflict clauses can therefore be used very effectively. In contrast, DynamicSAT Hybrid uses more structural information and generates less powerful conflict clauses. Therefore, more clauses are learned.

The impact on the set of aborted faults is enormous. Ninety-five percent of those faults on which PASSAT aborted (line *None*) were classified by PASSAT+ (line *All*). The number is even higher for DynamicSAT+ Hybrid. Here, 97% of all previously aborted faults were classified. DynamicSAT+ Hybrid also produces the fewest aborts in total. However, this strategy suffers from the large memory consumption. The run time behavior is very poor for those circuits where many conflicts are detected. This is due to the large number of reused clauses which have to be processed during BCP.

The learning strategies are able to reduce the memory needs of DynamicSAT+ Hybrid by reducing the amount of learned information. Strategy *Size 10* turned out to be most effective in reducing the aborts for DynamicSAT+ Hybrid. The problem of the large memory consumption is not given for PASSAT+, because PASSAT+ with learning strategy *All* does not learn as many pervasive clauses as DynamicSAT+ Hybrid does. Therefore, the problem of the large run time overhead is not as serious as for DynamicSAT+ Hybrid. However, the total number of unclassified faults of learning strategy *Act 1/2* is even smaller than of *All*. Therefore, this strategy is preferred.

The detailed experimental results for those strategies which turned out to be best are given in Table 6.5. Here, column *Imp.P* and column *Imp.D* shows the factor of improvement compared to PASSAT and DynamicSAT Hybrid, respectively (without dynamic learning). Column *Imp.A* gives the run time improvement compared to the respective approach using learning strategy *All.*

Table 6.5 Experimental results – dynamic learning

	PASSAT+ Act *1/2*					DynamicSAT+ Hybrid *Size 10*				
Circ.	Mem.	Ab.	Time	Imp.P	Imp.A	Mem.	Ab.	Time	Imp.D	Imp.A
b14	1.00x	0	0:42 min	0.93x	0.88x	1.02x	0	0:13 min	1.00x	1.00x
b15	1.00x	0	0:54 min	1.06x	1.15x	1.07x	0	2:18 min	0.49x	1.09x
b17	1.00x	0	2:34 min	0.97x	1.01x	1.09x	0	2:46 min	0.84x	1.08x
b18	1.09x	0	11:00 min	0.65x	0.81x	1.06x	0	6:13 min	1.11x	1.04x
p44k	0.99x	0	21:11 min	3.45x	0.98x	1.01x	0	5:20 min	0.97x	1.01x
p49k	1.80x	4	1:18 h	5.51x	1.04x	1.32x	0	5:47 h	2.25x	6.22x
p57k	1.02x	4	2:55 min	0.99x	1.08x	1.03x	3	1:00 min	0.97x	1.02x
p77k	1.00x	0	0:13 min	0.62x	1.00x	1.00x	0	0:08 min	1.00x	1.00x
p80k	1.01x	0	8:25 min	0.25x	0.96x	1.01x	0	1:25 min	1.01x	1.01x
p88k	1.00x	0	3:18 min	0.57x	1.11x	1.02x	0	1:32 min	0.99x	1.00x
p99k	1.03x	6	1:57 min	0.56x	0.98x	1.04x	8	1:23 min	0.92x	1.08x
p177k	1.01x	0	31:41 min	2.78x	0.96x	1.02x	0	6:23 min	0.93x	1.02x
p456k	1.10x	141	1:03 h	0.38x	1.22x	1.07x	204	1:32 h	0.64x	1.89x
p462k	1.00x	129	1:05 h	0.62x	0.98x	1.03x	61	1:19 h	0.45x	1.00x
p565k	1.00x	0	27:15 min	0.30x	1.00x	0.86x	0	17:59 min	1.00x	1.00x
p1330k	1.00x	0	49:39 min	0.64x	1.00x	1.02x	33	57:56 min	1.02x	1.10x
p2787k	1.00x	2	15:00 h	0.89x	0.99x	1.02x	36	6:51 h	0.99x	0.98x
p3327k	1.04x	425	33:26 h	0.60x	1.08x	1.03x	440	20:18 h	0.90x	1.12x
p3852k	1.02x	250	22:20 h	0.48x	0.94x	1.02x	224	12:51 h	1.18x	1.09x
Av./total	1.06x	961				1.04x	1,009			

Both approaches are very effective in reducing the unclassified faults for each circuit. The effects of both strategies are comparable. Generally, the run time is comparable or increases due to the amount of learned information which has to be processed. PASSAT+ is able to speed up test generation for some circuits, i.e. p44k, p49k and p177k, by several factors. These are those circuits where the run time of PASSAT is very high. However, PASSAT+ still needs more run time for these circuits than DynamicSAT+ Hybrid (except for p49k). DynamicSAT+ Hybrid reduces the run time for p49k significantly and is the only approach which is able to fully classify this circuit. The run time behavior of DynamicSAT+ Hybrid is generally better than that of PASSAT+. Therefore, DynamicSAT+ Hybrid using strategy *Size 10* is the first choice.

6.5.3.2 Post-Classification Phase

This section presents the experimental results of the Post-Classification Phase (PCP). However, only the best learning strategies are evaluated, i.e. PASSAT+ with *Act 1/2* and DynamicSAT+ Hybrid with *Size 10*. After performing the normal test generation process, the PCP is started trying to re-classify the aborted faults with all learned information available. Table 6.6 presents the concrete results. Column *Ab. DL* shows the number of aborted faults before starting the PCP and column

Table 6.6 Experimental results – post-classification phase

	PASSAT+ *Act 1/2*				DynamicSAT+ Hybrid *Size 10*			
Circ.	Ab. DL	Ab. PCP	Diff. (%)	Time (%)	Ab. DL	Ab. PCP	Diff. (%)	Time (%)
p49k	4	2	−50	+14	–	–	–	–
p57k	4	4	−0	+4	3	2	−33	+14
p99k	6	2	−66	+9	8	1	−88	+13
p456k	141	64	−55	+4	204	151	−26	+3
p462k	129	88	−32	+3	61	45	−26	+3
p1330k	–	–	–	–	33	29	−12	+4
p2787k	2	0	−100	+3	36	11	−69	+7
p3327k	425	206	−52	+2	440	290	−34	+4
p3852k	250	151	−40	+3	224	151	−33	+6
av./total	961	517	−49	+5	1,009	680	−40	+7

Ab. PCP shows the number of aborted faults after performing the PCP. The reduction of the aborts is given in percentage in column *Diff.* and the run time overhead for the PCP is shown in column *Time*.

Although the presented learning strategies are very effective in reducing the number of aborts, the PCP is able to further reduce the remaining unclassified faults by 49% for PASSAT+ *1/2 Act* and by 40% for DynamicSAT+ Hybrid *Size 10* on average. This shows that the information learned after aborting some faults can be effectively applied when targeting the fault again. The run time overhead for the PCP is mostly negligible being 5% and 7%, respectively, of the overall run time on average. Only for some small circuits, the overhead is higher (up to 14%). In summary, if the number of unclassified faults is too high after test generation, the PCP can be applied to further reduce the aborted faults with moderate run time overhead.

6.5.4 Combination of Structural and SAT-Based Algorithms

As described in Sect. 4.3, a structural and a SAT-based algorithm can be combined such that the advantages of both engines can be exploited. This section shows the results of such a combination, i.e. an industrial FAN-based engine and the proposed SAT-based engines. The results are presented in Table 6.7. Column *PASSAT* shows the results of the combination with the original PASSAT as proposed in [DEF+08]. Column *DynSAT+ Hybrid* presents the results for the combination of the FAN-based engine with the proposed approach DynamicSAT+ Hybrid using learning strategy *Size 10*. The column *Imp.* shows the improvement of the approach compared to the classical combination FAN + PASSAT.

The SAT-based algorithm is in most cases only applied to a small part of all faults (column *Calls*), since the majority of all faults can be classified by the structural algorithm. The run time degrades slightly for most circuits or displays a small

Table 6.7 Experimental results – combination

	FAN combined with						
	PASSAT			DynamicSAT+ Hybrid			
Circ.	Ab.	Time	Calls (%)	Ab.	Time	Imp.	Calls (%)
b14	0	0:27 min	0.7	0	0:26 min	1.04x	0.5
b15	0	0:53 min	18.9	0	1:17 min	0.69x	18.1
b17	0	2:33 min	18.0	0	2:55 min	0.87x	18.1
b18	0	5:08 min	3.9	0	5:03 min	1.02x	3.9
p44k	0	1:12 min	0.0	0	1:12 min	1.00x	0.0
p49k	6,562	5:03 h	46.3	27	4:28 h	1.13x	45.6
p57k	2	1:08 min	3.1	6	1:09 min	0.99x	3.0
p77k	0	0:07 min	0.0	0	0:07 min	1.00x	0.0
p80k	0	1:40 min	0.4	0	1:41 min	0.99x	0.5
p88k	0	1:32 min	0.5	0	1:35 min	0.97x	0.5
p99k	2	1:21 min	14.9	8	1:41 min	0.80x	15.0
p177k	0	2:24 min	0.9	0	2:10 min	1.11x	1.0
p456k	570	20:06 min	13.7	441	1:11h	0.28x	13.9
p462k	4	9:20 min	1.0	6	9:08 min	1.02x	1.1
p565k	0	8:28 min	4.0	0	8:59 min	0.94x	4.2
p1330k	0	12:34 min	2.1	1	13:04 min	0.96x	2.1
p2787k	4,153	10:52 h	20.6	13	3:19 h	3.28x	22.6
p3327k	1,784	6:12 h	6.2	374	7:04 h	0.88x	6.2
p3852k	1,733	10:10 h	6.5	308	11:11 h	0.91x	6.3
Av./total	14,808		8.5	1,184			8.6

improvement. A significant run time improvement could be achieved for circuit p2787k. In contrast, the run time becomes worse for circuit p456k.

Generally, the classical combination is able to completely (or almost completely) classify most circuits very quickly. However, a larger number of unclassified faults remains for a few circuits. The combination of FAN with DynamicSAT+ Hybrid is able to decrease the number of unclassified faults although the learning technique does not have the possibility to reuse as much recorded conflict clauses as when it is applied to all faults. However, the number is still slightly higher compared to the stand-alone application of SAT-based ATPG.

6.6 Summary

SAT-based algorithms generate a large number of conflict clauses during their search for a test pattern. However, the same parts of the circuits are often targeted since a large number of faults is targeted during an ATPG run. A subset of the learned information – the pervasive conflict clauses – can be extracted and reused for subsequent faults to prune large parts of the search space.

An efficient method to detect and reuse pervasive conflict clauses has been presented in this chapter. The use of a *Learned Watch List* (LWL) allows for an efficient embedding of the pervasive conflict clauses into subsequent ATPG calls. Furthermore, the LWL allows for a tight integration into the DCA technique presented in Chap. 5. Additionally, strategies to reduce the amount of learned information are presented and the concept of a *Post-Classification Phase* (PCP) is proposed which targets aborted faults again with more learned information available.

The experimental results clearly show the advantages of the resulting SAT-based ATPG engine. A large percentage of the aborted faults can be classified using the proposed techniques. At the same time, the run time overhead is very low. For some circuits, there is even a speed-up. Compared to a highly optimized industrial structural ATPG algorithm, the proposed techniques are able to reduce the number of aborts from over 200,000 to under 1,000. Besides the benefits as a stand-alone ATPG engine, the benefits of the proposed techniques have been shown in combination with a structural ATPG engine.

In summary, the integration of DCA and the proposed dynamic learning technique into the SAT-based engine results in a fast and highly robust SAT-based test generation process producing very few aborts.

Part III
High Quality Delay Test Generation

Chapter 7
High Quality ATPG for Transition Faults

The *Transition Fault Model* (TFM) takes the prevalent position among the delay fault models in the industrial production test. This fault model is widely used to ensure that a manufactured circuit is free of delay defects. One major reason behind the widespread use of the TFM is the similarity to the SAFM. This allows for the application of existing ATPG algorithms, i.e. only one ATPG engine has to be maintained for both fault models. Furthermore, the TFM has a small fault population compared to the PDFM. The number of faults is linear in the number of gates. Designs which have a good stuck-at fault coverage typically also have a good transition fault coverage [WLRI87]. However, the growing complexity and size of today's designs lead to a serious problem in industrial practice. Classical structural ATPG algorithms produce a large number of aborts and compromise the high fault coverage demands of the industry. The advantages of using SAT-based algorithms for test generation for the SAFM have already been shown in the previous chapters.

This chapter deals with the SAT-based test generation for the TFM. Section 7.1 presents the SAT formulation for the TFM and shows experimental results on large industrial circuits using a classical SAT engine as well as the SAT techniques presented in Chaps. 5 and 6. In particular, the results show that the resulting SAT-based ATPG framework is highly fault-efficient leading to a significant increase of the transition fault coverage of up to 2%.

Another aspect of test generation for delay faults is the test quality. Due to the shrinking feature sizes and process variations, the likelihood of small delay defects increases and becomes a serious issue in industrial practice [KC98]. Unfortunately, the TFM does not have the inherent ability for the detection of small delay defects, because it assumes a lumped fault effect to be large enough to be observed at any output. Section 7.2 presents an efficient SAT instance generation technique that prioritizes long propagation paths and, therefore, produces tests that are more likely to detect small delay defects. As a result, test patterns of higher quality are generated in a very scalable manner.

Timing-aware ATPG promises an even higher test quality since the fault is guaranteed to be activated and propagated via the longest path. However, the

S. Eggersglüß and R. Drechsler, *High Quality Test Pattern Generation and Boolean Satisfiability*, DOI 10.1007/978-1-4419-9976-4_7,

existing approaches for timing-aware ATPG are typically based on structural ATPG techniques and consequently share their disadvantages. These approaches typically decrease the accuracy in order to cope with the high complexity of this problem. Section 7.3 presents a pseudo-Boolean formulation for timing-aware ATPG which strongly relies on the efficient SAT techniques and is able to model transition-dependent delays. Section 7.4 summarizes this chapter.

7.1 Transition Fault Model: SAT Formulation

This section describes the TF formulation in CNF [ETF+07]. Section 7.1.1 shows how two time frames are handled in CNF and Sect. 7.1.2 describes the SAT formulation for fault modeling.

7.1.1 Iterative Logic Array

Since two time frames are needed to generate a transition test (see Sect. 2.3.2), the SAT instance for transition test generation has to incorporate the behavior of the circuit in these time frames. The work presented in [KL93] models the sequential behavior of the circuit as an *Iterative Logic Array* (ILA). The circuit's sequential behavior is mapped onto a combinational circuit by unrolling the combinational logic k times.

A length of $k = 2$ is needed for transition test generation. The first time frame t_1 is called the initial time frame. The second time frame t_2 is called final time frame. In the following, the SAT formulation of such an ILA is described in detail. Here, the launch-on-capture test scheme [SP94] is used.

Since the combinational part has to be unrolled two times, the combinational part of circuit $\mathscr{C}$ is duplicated resulting in two circuits $\mathscr{C}_1$ and $\mathscr{C}_2$. Circuit $\mathscr{C}_1$ represents t_1 and circuit $\mathscr{C}_2$ represents t_2. Today's circuits are typically sequential circuits, i.e. they contain state elements like flip-flops. Therefore, these elements have to be modeled, too. Only the initial value can be scanned into a state element in a standard scan design using launch-on-capture (*current state*). The final value of a state element is calculated by the combinational logic during t_1 (*next state*). Therefore, state elements are modeled by connections between the state elements in $\mathscr{C}_1$ and their counterparts in $\mathscr{C}_2$. Typically, a buffer is modeled to ensure the integrity of the signal.

In order to apply a SAT solver, the unrolled circuit $\mathscr{C}_t$ must be transformed into a CNF derived from the following equation:

$$\Phi_{\mathscr{C}_t} = \Phi_{\mathscr{C}_1} \cdot \Phi_{\mathscr{C}_2} \cdot \Phi_{\text{seq}}$$

The CNF for $\mathscr{C}_1$ is represented by $\Phi_{\mathscr{C}_1}$ and $\Phi_{\mathscr{C}_2}$ is the CNF for $\mathscr{C}_2$. The circuit-to-CNF transformation was described in Sect. 3.4.

The term Φ_{seq} describes the sequential behavior of the unrolled circuit $\mathscr{C}_t$, i.e. the modeling of state elements, in a standard scan design. By omitting Φ_{seq}, the CNF would represent a combinational circuit or an enhanced scan design. The concrete fault modeling is described in the next section.

An additional constraint must be considered for the use in industrial practice. The value of the PIs (not PPIs) cannot change during t_1 and t_2 and has to be equivalent. This is due to the test equipment which is typically not able to change the test value on the PIs at speed during the test application.

7.1.2 Injection of Stuck-at Faults

The modeling of TFs for SAT-based ATPG in an ILA of length $k = 2$ is discussed in detail in this section. TFs can be modeled by injecting stuck-at faults with the circuit modeled in two consecutive time frames as shown in [LM86, WLRI87]. The initial time frame t_1 is used to initiate the transition and the transition is launched and propagated to an output in the final time frame t_2.

In order to initialize the transition, the faulty line f is fixed to the initial value of the transition in t_1, i.e. 0 for a slow-to-rise fault $(f, \uparrow)$ and 1 for a slow-to-fall fault $(f, \downarrow)$. This constraint is described by Φ_{F1} in the following. Then, a stuck-at fault is injected at f in t_2 to launch and propagate the transition. Here, a stuck-at-0 fault is injected for $(f, \uparrow)$ and a stuck-at-1 fault is injected for $(f, \downarrow)$. Note that the same SAT formulation as described in Chap. 4 can be used for the stuck-at fault modeling in t_2 (see Fig. 4.3). However, the CNF for the stuck-at modeling consisting of the fault excitation condition, the faulty cone and the D-chain encoding (see Sect. 4.1.2) is derived only from $\mathscr{C}_2$. These constraints are merged by Φ_{F2}. In summary, the SAT instance $\Phi_{\text{test}}^{\text{TF}}$ for generating a test for a TF is given by:

$$\Phi_{\text{test}}^{\text{TF}} = \Phi_{\mathscr{C}_t} \cdot \Phi_{F1} \cdot \Phi_{F2}$$

Note that likewise to the SAT instance generation for the SAFM, $\Phi_{\mathscr{C}_t}$ only has to include the relevant parts of the circuits. This includes the fanin cone of the faulty line in t_1 and the fanin cones of all outputs to which the fault can be structurally propagated (spread over both time frames t_1 and t_2). The following example demonstrates the presented SAT instance generation for the TFM.

Example 7.1. Figure 7.1 shows an example circuit in its original sequential form with a slow-to-fall TF at line b. Note that for reason of simplicity, the circuit does not have an output, but one state element. The unrolled circuit is presented in Fig. 7.2. Both circuits $\mathscr{C}_1$ and $\mathscr{C}_2$ are concatenated by a connection replacing the flip-flop FF. Since line b corresponds to the faulty line in the initial time frame, b is constrained to hold the value 1 to initialize the TF. In order to launch and propagate the transition, a s-a-1 fault has to be injected at line b_2 which corresponds to the faulty line in the final time frame. The CNF $\Phi_{\mathscr{C}_t}$ for the unrolled circuit is given by the conjunction of the following CNFs $\Phi_{\mathscr{C}_1}$, $\Phi_{\mathscr{C}_2}$ and Φ_{seq}:

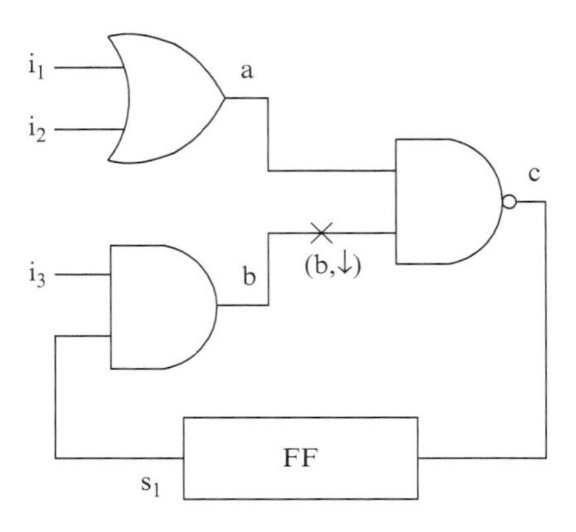

Fig. 7.1 Example circuit with TF

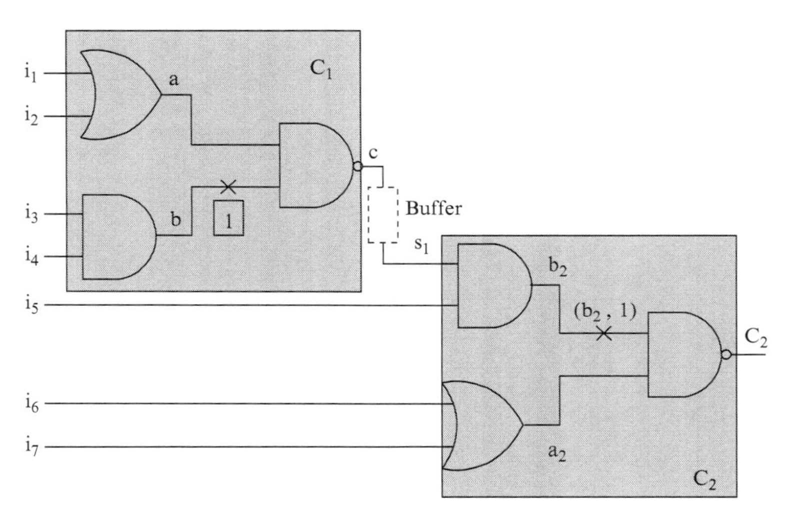

Fig. 7.2 Unrolled example circuit with stuck-at fault injection

$$\Phi_{\mathscr{C}_1} = \Phi_a^{\text{OR}} \cdot \Phi_b^{\text{AND}} \cdot \Phi_c^{\text{NAND}}$$
$$\Phi_{\mathscr{C}_2} = \Phi_{a_2}^{\text{OR}} \cdot \Phi_{b_2}^{\text{AND}} \cdot \Phi_{c_2}^{\text{NAND}}$$
$$\Phi_{\text{seq}} = (c + \overline{s}_1) \cdot (\overline{c} + s_1)$$

where the term $\Phi_{\text{line}}^{\text{gatetype}}$ denotes the CNF for the particular gate type and the particular line. The constraints for modeling the slow-to-fall TF on line b are given by the following equations (see Sect. 4.1.2 for more information about fault modeling and D-chain encoding):

$$\Phi_{f1} = (b)$$
$$\Phi_{f2} = \Phi_{c_2^f}^{\text{NAND}} \cdot \Phi_d \cdot (b_2) \cdot (c_2^d)$$

Here, the line c_2^f belongs to the faulty cone, Φ_d represents the D-chain encoding and the remaining unit clauses are used for fault excitation. A corresponding test for the slow-to-fall TF on line b obtained by the evaluation of $\Phi_{\text{Test}}^{\text{TF}}$ is:

$$V_1 = \{i_1 = 1, i_2 = 0, i_3 = 1, i_4 = 0\}$$
$$V_2 = \{i_5 = 1, i_6 = 1, i_7 = 1\}$$

As described above, an additional constraint must be added in practice. The values of a PI (not PPI) must be equivalent in both time frames. This constraint is incorporated by the following CNF Φ_{eq}:

$$\Phi_{\text{eq}} = (\bar{i}_1 + i_6) \cdot (i_1 + \bar{i}_6) \cdot (\bar{i}_2 + i_7) \cdot (i_2 + \bar{i}_7) \cdot (\bar{i}_3 + i_5) \cdot (i_3 + \bar{i}_5)$$

The test presented above is therefore invalid, because i_2 and i_7 are not equal. A valid test satisfying the constraint Φ_{eq} is:

$$V_1 = \{i_1 = 1, i_2 = 1, i_3 = 1, i_4 = 0\}$$
$$V_2 = \{i_5 = 1, i_6 = 1, i_7 = 1\}$$

7.1.3 Experimental Results

The experimental results for test generation for the TFM are presented in this section. Note that the same experimental setup as for the SAFM described in Sect. 5.7.3 is used. First, Table 7.1 shows a comparison of the structural ATPG algorithms FAN and FAN long (with increased resources) with the PASSAT approach.[1] The last row of each table gives the total number of aborts.[2] If the general ATPG process was aborted due to the timeout limit of 72 CPU hours, the percentage of targeted faults in relation to the total number of faults is given instead of the run time.

[1] Note that no results for circuit p49k are given since nearly all targeted faults remain unclassified by all considered approaches. The results for this circuit are therefore not meaningful.

[2] Faults not targeted during ATPG due to the timeout are not included.

Table 7.1 Experimental results – transition faults – FAN/PASSAT

	FAN		FAN long		PASSAT		
Circ.	Ab.	Time	Ab.	Time	Ab.	Time	Imp.F
b14	417	3:21 min	219	14:02 min	0	2:59 min	1.12x
b15	3,579	10:17 min	2,642	1:16 h	0	3:55 min	2.63x
b17	13,977	45:42 min	9,220	6:04 h	0	17:25 min	2.62x
b18	43,781	5:14 h	26,970	20:36 h	0	1:53 h	2.78x
p44k	648	18:45 min	539	30:29 min	0	5:38 h	0.06x
p57k	2,467	14:45 min	1,465	1:06 h	44	1:00 h	0.25x
p77k	49,452	9:05 min	36,803	1:49 h	5,421	1:17 h	0.12x
p80k	4,527	10:56 min	3,022	53:35 min	9	12:15 min	0.89x
p88k	6,964	1:13 h	3,833	3:54 h	0	33:44 min	2.16x
p99k	13,448	56:14 min	11,462	8:28 h	73	21:55 min	2.57x
p177k	13,206	46:53 min	7,992	3:25 h	19,002	*97.0%*	–
p456k	64,568	3:50 h	52,219	21:36 h	29,693	17:50 h	0.21x
p462k	17,452	1:15 h	13,353	3:41 h	551	7:09 h	0.17x
p565k	9,708	3:47 h	5,548	7:37 h	412	4:12 h	0.90x
p1330k	11,328	2:40 h	5,679	4:12 h	3	9:18 h	0.29x
p2787k	698,387	19:22 h	321,025	*16.8%*	34,689	*70.0%*	–
p3327k	250,547	*94.3%*	60,159	*99.5%*	16,694	*68.1%*	–
p3852k	88,091	*90.2%*	54,314	*84.6%*	1,353	*63.1%*	–
Total	1,292,547		616,464		107,944		

The results lead to similar observations as for the SAFM. Structural ATPG is fast and produces many aborts, i.e. over a million unclassified faults in total in the experiments conducted. Generally, the results of FAN show that the problem of the high number of unclassified faults which already exists for the SAFM also exists for the TFM. However, the problem is even more serious. Increasing the resources yields a reduction in terms of aborts but results in significant run time overhead. In contrast, PASSAT is much slower than FAN in most cases but produces significantly less aborts (even less than FAN long). A difference can be observed between the benchmark circuits and the industrial circuits. Here, PASSAT is constantly faster than FAN.

Applying the DCA technique leads to similar effects as for the SAFM as well. The SAT-based test generation process is significantly accelerated as reported in Chap. 5. Table 7.2 gives the results of the proposed approaches using DCA in combination with circuit-based dynamic learning, i.e. DynamicSAT+ and DynamicSAT+ Hybrid. Here, the *Size 20* learning strategy turned out to be most effective in preliminary evaluations. Learning strategies with a clause size larger than 20 improve the number of aborts only slightly but increase the run time considerably. Therefore, this strategy is used for the TFM. The factor of improvement compared to FAN (PASSAT) is given in column *Imp.F* (*Imp.P*).

The results show the advantages of the proposed techniques. Very few faults remain unclassified compared to the other approaches, i.e. FAN, FAN long and PASSAT. Furthermore, the approaches are able to speed up SAT-based test generation

Table 7.2 Experimental results – transition faults – DynamicSAT+ (Hybrid)

	DynamicSAT+				DynamicSAT+ Hybrid			
Circ.	Ab.	Time	Imp.P	Imp.F	Ab.	Time	Imp.P	Imp.F
b14	3	2:27 min	1.22x	1.37x	1	2:45 min	1.08x	1.22x
b15	4	21:23 min	0.18x	0.48x	0	21:22 min	0.18x	0.48x
b17	6	58:03 min	0.30x	0.79x	0	59:18 min	0.29x	0.77x
b18	745	3:25 h	0.55x	1.53x	7	3:20 h	0.57x	1.57x
p44k	42	19:34 min	17.27x	0.96x	37	22:23 min	15.10x	0.84x
p57k	74	15:08 min	3.96x	0.97x	36	19:08 min	3.14x	0.77x
p77k	0	5:42 min	13.51x	1.59x	111	7:29 min	8.19x	1.21x
p80k	5	8:44 min	1.40x	1.25x	7	9:01 min	1.36x	1.21x
p88k	0	35:45 min	0.94x	2.04x	0	35:55 min	0.94x	2.03x
p99k	74	30:12 min	0.73x	1.86x	35	29:25 min	0.75x	1.91x
p177k	277	2:12 h	32.73x	0.36x	483	13:44 h	3.15x	0.06x
p456k	9,768	47:10 h	0.38x	0.08x	5,955	61:21 h	0.29x	0.06x
p462k	19	1:14 h	5.80x	1.01x	6	1:39 h	4.33x	0.76x
p565k	355	2:44 h	1.54x	1.38x	133	3:01 h	1.39x	1.25x
p1330k	15	2:47 h	3.34x	0.96x	6	2:44 h	3.40x	0.98x
p2787k	5,886	*68.9%*	–	–	4,370	*63.1%*	–	–
p3327k	9,069	*77.8%*	–	–	3,119	*77.5%*	–	–
p3852k	7,913	*84.9%*	–	–	3,226	*85.1%*	–	–
Total	34,255				17,532			

significantly for those circuits where PASSAT is slow. The highest speed-up factor is 32.72 for p177k, where PASSAT is not able to finish test generation within 72 CPU hours. However, the only industrial circuit, where the run time results are not satisfactory is p456k, where PASSAT produces many aborts. The run time of DynamicSAT+ (Hybrid) increases due to the large amount of learned information. However, the number of unclassified faults are significantly reduced in return.

Table 7.3 shows the impact of the proposed SAT-based ATPG approach DynamicSAT+ Hybrid on the fault coverage and fault efficiency for the TFM compared to the industrial FAN-based ATPG. Column *%FC* gives the fault coverage which could be achieved with the corresponding approach and column *%FE* presents the fault efficiency. The fault efficiency (or test coverage)[3] is defined as the percentage of testable faults in faults not identified as untestable. Column *%FC Inc.* gives the fault coverage increase of DynamicSAT+ compared to FAN.

The fault efficiency of DynamicSAT+ is very high being either 100% or between 99.5% and 100%. This signifies a considerable increase compared to FAN and shows the robustness of the approach. Furthermore, the application of DynamicSAT+ results in a significant fault coverage increase of up to 2% which is very important for the high quality demands of today's designs.

[3]Fault efficiency is usually taken as a measurement for the effectiveness of the ATPG engine.

Table 7.3 Impact on fault coverage/fault efficiency

	FAN		DynamicSAT+ Hybrid		
Circ.	%FC	%FE	%FC	%FE	%FC Inc.
p44k	55.15	99.40	55.36	99.98	+0.21
p57k	96.36	98.71	97.23	99.99	+0.87
p77k	34.46	67.62	34.46	99.92	+0.00
p80k	94.86	98.58	96.06	100.00	+1.20
p88k	92.33	97.56	94.00	100.00	+1.67
p99k	89.91	95.95	90.91	99.99	+1.00
p177k	76.13	96.56	77.54	99.91	+1.41
p456k	84.17	94.43	86.18	99.50	+2.01
p462k	57.68	97.48	57.95	100.00	+0.27
p565k	94.81	99.44	95.02	100.00	+0.21
p1330k	90.44	99.54	90.57	100.00	+0.13

7.2 Long Propagation Paths

A *Small Delay Defect* (SDD) is a defect with defect size not large enough to cause a timing failure by its own. Due to the shrinking feature sizes and the increased speed of today's circuits, the likelihood of failures caused by SDDs increases and their detection has become an important issue in the production test [KMGE04]. Although being very small, SDDs might cause a timing violation when many of them are accumulated. This is also called a distributed delay defect.

An SDD might escape during test application when a short path is sensitized since the accumulated delay of the distributed delay defect is not large enough to cause a timing violation. In contrast, the same SDD might be detected if a long path is sensitized [GH04, KMGE04]. Unfortunately, common ATPG algorithms usually prefer short paths for fault propagation since the sensitization of these paths is typically easier. In the following, an overview on approaches is given which were developed in order to enhance the test quality by increasing the probability of detecting SDDs.

A PODEM-based algorithm for detecting high quality tests for transition faults was proposed in [SPR02]. But this algorithm focuses on the predetermination of path sensitization conditions for the longest paths. An approach for testing SDDs based on the TF model is presented in [PG06]. There, standard TF testing is combined with information gathered from static timing analysis. By grouping test patterns and adjusting their timing, the paths can be tested almost with no slack and by this obtain a higher coverage of small delay defects.

A new TF model called *As Late As Possible Transition Fault* (ALAPTF) was proposed in [GH04] that tries to activate the fault as late as possible and sensitizes a long path to achieve a good detection rate of small delay defects. However, the ATPG method is computationally complex due to the timing information used. Timing-aware ATPG was proposed in [LTW+06]. Here, timing information is leveraged during the search for a test in order to sensitize the longest path. However, this is also a very time-consuming task as reported in [YCT08]. Timing-aware ATPG is discussed in more detail in Sect. 7.3.

In [ATJ06], an ATPG algorithm for detecting small delay defects that is based on path delay testing was introduced. The algorithm generates a set of multiple-detect test patterns and uses a pattern selection strategy to sensitize the long paths rather than the short paths. In [YCT08], a test pattern grading technique was proposed that selects test patterns from an n-detection pattern set according to their effect on small delay defect detection. Here, the test is graded by its output deviation [WC08].

This section presents a SAT technique which prioritizes long propagation paths during test generation [TED10]. By this, the quality of the test set is increased, i.e. the likelihood of detecting distributed delay defects.

Unfortunately, information about the circuit and especially about its timing is lost during the SAT instance generation. Therefore, the SAT instance generation is modified to guide the search towards sensitizing a long propagation path in a very scalable manner. In contrast to previous approaches, the timing information is leveraged in order to incrementally build the SAT instance and it is not used directly during the search. This is advantageous since the efficiency of today's SAT solvers is partly based on the homogeneous structure of the CNF that allows for fast BCP.

The proposed procedure is based on incremental SAT instance generation which is described in Sect. 7.2.1. A new output ordering scheme for prioritizing long propagation paths is given in Sect. 7.2.2 and experimental results are given in Sect. 7.2.3.

7.2.1 Incremental Instance Generation

Typically, the SAT instance generation is performed by traversing all transitive fanin cones of each output in the output cone of the fault site. However, showing the fault effect at one output only is sufficient to detect the fault. Therefore, incremental SAT instance generation was proposed in order to accelerate both the SAT instance generation as well as the solving time. In this book, a modification of this technique is shown with the aim to increase the quality of the test set for TFs.

Figure 7.3 shows the flow of this technique. At first, the initial SAT instance Φ_J is built. The initial SAT instance Φ_J consists of the fault site and its transitive fanin cone including the modeling of the fault. Then, a PO o_1 of the output cone of the fault site is chosen and the CNF Φ_{o_1} of the transitive fanin cone of o_1 is added to the initial SAT instance.

$$\Phi_1 = \Phi_J \cdot \Phi_{o_1}$$

If $\Phi_1 = 1$, i.e. satisfiable, a test is found. Otherwise, no classification can be given, since the fault might be propagated to another output.

Therefore, Φ_1 is augmented incrementally by the transitive fanin cone of the second output o_2. Note that only that part has to be added which is not included in

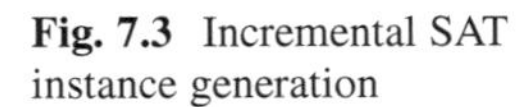
Fig. 7.3 Incremental SAT instance generation

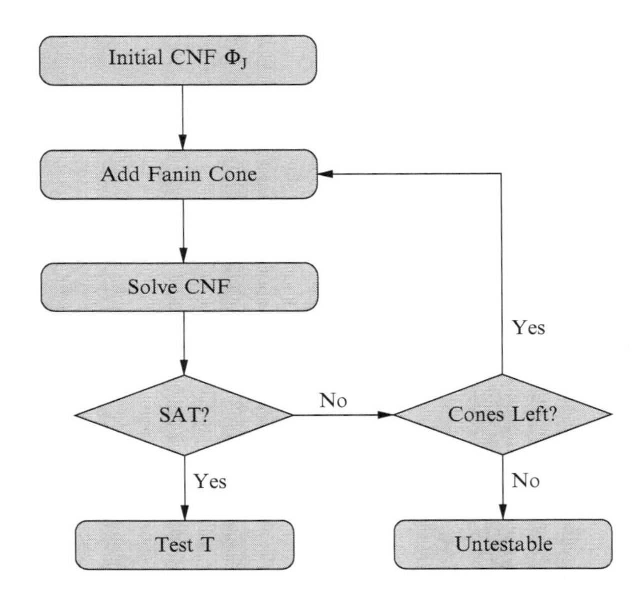

Φ_1. The conflict clauses recorded so far are kept. By this, the search process benefits because the already explored search space does not have to be traversed again. This procedure continues until either a test T is found or no more outputs are left. In this case, the fault is untestable.

7.2.2 *Output Ordering*

The efficiency of the procedure strongly depends on the output ordering. Therefore, two different output ordering schemes were proposed in [TED10] in order to improve the performance:

- Shortest path – the outputs are ordered with respect to the length to the fault site in gates. The nearest outputs are processed first since a short path is typically easier to sensitize.
- Smallest CNF – the outputs are ordered with respect to the CNF size of the transitive fanin cone. Those outputs with a smaller CNF are processed first, since a smaller CNF is typically easier to solve than a larger CNF.

As mentioned above, generating long propagation paths is generally preferred for detecting small delay defects. The direct use of timing information in order to find the longest sensitizable path results in a highly increased run time as described above. Instead, the use of timing information during the SAT instance generation is proposed. Therefore, the output ordering scheme is enriched with timing information:

- Longest path – The outputs are ordered with respect to their "distance" to the fault site. However, in contrast to the *shortest path* scheme, those outputs with a long path to the fault site are processed first.

Note that finding the longest propagation path is not guaranteed since there could be more than one path from the fault site to the respective output. Since the efficient search process is not altered, the path is chosen heuristically.

Due to the absence of a timing simulator, the current implementation is based on the fixed delay model where each gate in the circuit has a fixed delay of 1. The simple assumption is easily extensible by technology-dependent parameters. Since the ordering of the outputs is done prior to the solving step, the integration would be straightforward.

However, a large number of outputs in the output cone of the fault site would result in significant overhead if the longest paths are not testable adding the fanin cone of each single output in one step. Therefore, an n-steps approach is presented. This approach restricts the number of incremental steps to a number $n \in \mathbb{N}$ and adds more than the fanin cone of one output in each single step.

The number of outputs added in one single step depends on n. For example, if $n = 4$, than 25% of the outputs on a potential D-chain are added in one single step. This results in at most four incremental steps. As the experiments in the next section will show, the n-steps approach speeds up the search process and at the same time degrades the length of the propagation path only slightly.

7.2.3 Experimental Results

The experimental results of the SAT-based approach prioritizing long propagation paths are presented in this section. All experiments were conducted with the DynamicSAT+ Hybrid approach using the *Size 20* learning strategy. Table 7.4 presents the experimental results for the incremental approach processing each output in one step. The results of the regular approach without prioritizing long propagation paths (taken from Table 7.2) are given for reason of comparison, too. Table 7.5 shows the results for n-steps approaches with $n = 4$, $n = 8$ and $n = 32$.

The number of aborts are given in columns named *Ab.* and the run time is given in columns *Time*. Besides the average length of the propagation path (given in column L_P), the average length of the activation path is also given in column L_A in order to show that the total length of the sensitized path is not decreased by the proposed technique.

Comparing the run time of the non-incremental and the incremental approach, it can be observed that the run time decreases slightly for some circuits. However, the run time is increased for the majority of circuits. But most circuits need only a small run time overhead. The highest run time overhead is produced at b15 with a factor of 2.4. Circuit p456k could not be solved completely and approximately 15% fewer faults of the two largest circuits p2787k, p3327k and p3852k could be

Table 7.4 Experimental results – long propagation paths

	Non-incremental				Incremental			
Circ.	Ab.	Time	L_A	L_P	Ab.	Time	L_A	L_P
b14	1	2:45 min	16.6	12.2	0	2:00 min	17.3	18.6
b15	0	21:22 min	15.5	8.1	0	50:30 min	16.2	15.7
b17	0	59:18 min	19.9	8.5	0	1:27 h	19.8	18.7
b18	7	3:20 h	28.8	10.6	13	2:44 h	28.3	17.5
p44k	37	22:23 min	9.7	14.5	27	30:50 min	9.8	17.1
p57k	36	19:08 min	9.5	9.2	78	24:23 min	9.7	20.4
p77k	111	7:29 min	2.5	2.5	198	7:37 min	2.5	2.5
p80k	7	9:01 min	13.5	9.1	20	12:46 min	13.3	11.3
p88k	0	35:55 min	9.5	11.6	0	36:38 min	9.7	21.9
p99k	35	29:25 min	12.1	14.4	38	34:10 min	11.9	20.5
p177k	483	13:44 h	8.4	11.1	220	18:12 h	6.8	18.2
p456k	5,955	61:21 h	14.9	21.4	8,514	*96.3%*	14.5	26.4
p462k	6	1:39 h	8.9	11.8	8	1:54 h	8.8	19.7
p565k	133	3:01 h	9.3	19.9	184	4:03 h	9.3	29.0
p1330k	6	2:44 h	8.7	15.6	297	4:24 h	9.6	22.2
p2787k	4,370	*63.1%*	10.8	14.4	3,526	*64.8%*	11.3	19.2
p3327k	3,119	*77.5%*	12.8	27.1	411	*62.5%*	9.6	20.2
p3852k	3,226	*85.1%*	11.9	23.9	211	*69.4%*	7.7	23.4
Av./total	17,532		12.4	13.7	13,745		12.0	19.0

processed. Despite the overall performance loss, the robustness of SAT-based ATPG is maintained. Only few aborts are produced.[4]

As an advantage, the average length of the propagation path is significantly increased by up to a factor of 2.2 (b17, p57k). On average, the length of the propagation path is increased from 13.7 to 19.0 which corresponds to a factor of 1.4. However, the average propagation path length is decreased for p3327k and p3852k. This is due to the different fault set which is targeted during test generation. Since the test generation of both circuits was not completed, less faults were targeted in the incremental approach. The results for the n-steps approaches presented in Table 7.5 show that if more faults can be targeted before the timeout limit is reached, the average length of the propagation goes up.

The results of the n-steps approaches shows the scalability of the approach. The run time behavior is rather balanced between the 4-steps approach and the non-incremental approach. Slightly more faults are aborted. But the average length of the propagation path already increases. The 8-steps approach even has a slight advantage in the performance compared to the non-incremental approach. But similar to the 4-steps approach, the number of aborts grows slightly. The average length of the propagation path increases further using 8-steps. The average length of

[4]In fact, the total number of aborts is less than the total number of the normal approach. However, this is caused by the fewer amount of faults processed at p3327k and p3852k.

Table 7.5 Experimental results – long propagation paths – *n*-steps

Circ.	4-steps Ab.	Time	L_A	L_P	8-steps Ab.	Time	L_A	L_P	32-steps Ab.	Time	L_A	L_P
b14	0	2:16 min	17.1	13.2	0	2:16 min	17.3	15.1	0	2:04 min	17.3	17.6
b15	0	20:17 min	16.1	7.1	0	21:37 min	15.9	9.0	0	25:48 min	15.9	13.9
b17	0	54:25 min	20.1	9.2	0	55:18 min	20.0	11.2	0	1:00 h	20.1	17.7
b18	12	3:22 h	28.5	12.4	243	3:33 h	28.6	13.0	215	3:30 h	27.9	16.4
p44k	36	23:39 min	9.8	14.9	24	21:46 min	9.8	15.3	38	24:31 min	9.7	16.8
p57k	79	19:55 min	9.4	11.0	63	20:08 min	9.5	12.1	61	19:36 min	9.4	16.3
p77k	196	8:40 min	2.5	2.5	202	8:27 min	2.5	2.5	198	8:54 min	2.5	2.5
p80k	11	10:10 min	13.5	10.1	20	10:38 min	13.5	10.6	5	9:40 min	13.5	11.1
p88k	0	35:26 min	9.6	14.7	0	35:12 min	9.6	16.8	0	35:48 min	9.6	18.5
p99k	38	32:26 min	11.9	18.2	42	34:06 min	11.6	20.2	37	34:17 min	11.7	20.5
p177k	554	15:34 h	8.6	16.8	483	9:30 h	8.5	17.2	409	14:37 h	8.5	17.6
p456k	7,271	*95.6%*	14.5	25.9	9,389	*97.3%*	14.3	25.5	8,214	*95.8%*	14.7	27.2
p462k	3	1:35 h	8.5	13.4	7	1:45 h	8.6	16.3	5	1:51 h	8.7	18.6
p565k	161	3:15 h	9.3	25.7	203	2:58 h	9.5	27.6	195	3:42 h	9.4	28.9
p1330k	210	2:58 h	9.7	17.0	283	3:05 h	9.7	17.6	278	3:19 h	9.5	22.5
p2787k	4,252	*65.4%*	11.4	17.3	4,080	*65.4%*	11.4	18.2	3,671	*65.3%*	11.3	18.9
p3327k	4,183	*80.1%*	13.1	27.1	3,864	*79.3%*	12.4	27.7	431	*64.4%*	9.5	21.3
p3852k	5,416	*84.9%*	11.4	21.6	4,507	*85.2%*	11.2	22.8	2,026	*78.2%*	11.3	27.0
Av./total	22,422		12.5	15.5	23,420		12.4	16.6	15,783		12.3	18.6

the propagation path of the 32-steps approach nearly reaches the high average length of the original incremental approach. However, the run time behavior is improved in most cases.

In summary, the incremental SAT instance generation is shown to be able to improve the quality of the test set in terms of path length. The technique is well scalable and the influence on the number of aborts is very low. That is, the high level of robustness of SAT-based ATPG is maintained and the performance is only slightly influenced.

7.3 Timing-Aware ATPG

This previous section showed a method how to increase the length of the propagation path in a very scalable manner. However, this method is based on a heuristic and therefore not exact. Although the average path length is increased, the approach might not find a long path for some faults. This is disadvantageous if the critical paths of the circuit are missed. Furthermore, the approach targets only long propagation paths, the length of the activation path is ignored during test generation.

This necessitates an exact method. Timing-aware ATPG is typically used to generate a test which propagates the fault via the longest path. However, timing-aware ATPG makes heavy use of timing information during the actual search process and is reported to have an significant run time overhead [YCT10] for larger circuits.

This section proposes a new approach [ED11c] for timing-aware ATPG which is based on *Pseudo-Boolean Optimization* (PBO). Modern PBO solvers are strongly influenced by SAT solving techniques and can leverage the efficient techniques of this domain.

Section 7.3.1 gives a motivational example of the timing-aware ATPG problem. Then, basic information about the PBO problem and its application to circuit-oriented problems are given in Sect. 7.3.2. Here, the similarities and the differences to the SAT problem are also briefly described. Section 7.3.3 shows how the timing-aware ATPG problem can be formulated as a PBO problem. The usage of structural information in the problem formulation is given in Sect. 7.3.4. Section 7.3.5 describes how more accurate transition-dependent delay values (instead of fixed delay values) are incorporated in the problem formulation. Experimental results are given in Sect. 7.3.6.

7.3.1 Motivational Example

Common ATPG algorithms tend to sensitize short paths during test generation due to reasons of complexity. However, this is disadvantageous for detecting SDDs. Delay

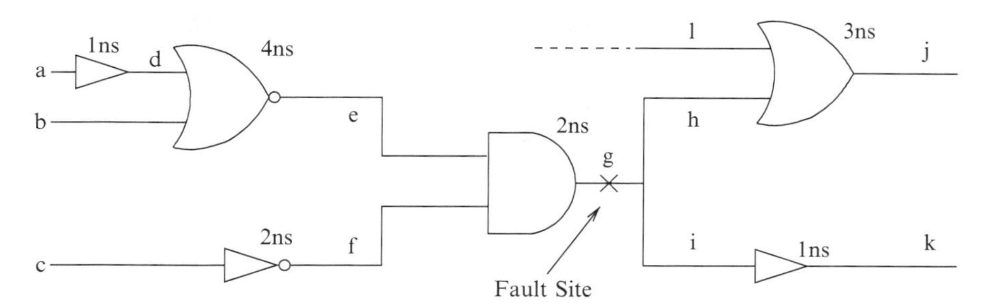

Fig. 7.4 Example circuit for timing-aware ATPG

defects based on SDDs are more likely to occur on longer paths, since more SDDs can be potentially accumulated and the slack margin is smaller. This is demonstrated by the following example.

Example 7.2. Consider the simple example circuit shown in Fig. 7.4. Each gate is associated with a specific delay. Assume that the fault site is line g. There are six possible paths through g on which the transition could be propagated:

- p_1 = a–d–e–g–h–j (10ns)
- p_2 = b–e–g–h–j (9ns)
- p_3 = a–d–e–g–i–k (8ns)
- p_4 = b–e–g–i–k (7ns)
- p_5 = c–f–g–h–j (7ns)
- p_6 = c–f–g–i–k (5ns)

Regular ATPG tools try to find a path, on which the transition is propagated, as fast as possible. So, it is most likely that a regular ATPG algorithm sensitizes the shortest path p_6, since this is the easiest path to sensitize. If the value is sampled for example at 11ns, the slack margin is very high, i.e. the accumulated defect size has to be at least 7ns for p_6 to detect a delay defect. However, if the ATPG algorithm chooses path p_1, the defect size has to be only 2ns for a detection.

Timing-aware ATPG [LTW+06] was developed to enhance the quality of the delay test. Here, a test is generated to detect the transition fault through the longest path by using timing information during the search. The algorithm proposed in [LTW+06] is based on structural ATPG and consists of two tasks: fault propagation and fault activation. Each task uses the path delay timing information as a heuristic to propagate (activate) the fault through the path with maximal static propagation delay (maximal static arrival time). However, due to complexity reasons, both tasks are carried out independently and the longest path might be missed. Furthermore, simplifications are assumed to further reduce the complexity. This motivates the need for new techniques that can cope with the high complexity.

Table 7.6 Pseudo-Boolean and CNF representation for an AND gate $a \cdot b = c$

PB	CNF
$((1-a)+(1-b)+c \geq 1)\cdot$	$(\overline{a}+\overline{b}+c)\cdot$
$(a+(1-c) \geq 1)\cdot$	$(a+\overline{c})\cdot$
$(b+(1-c) \geq 1)$	$(b+\overline{c})$

7.3.2 *Pseudo-Boolean Optimization*

This section gives basic information about *Pseudo Boolean Optimization* (PBO) and the related *Pseudo-Boolean* (PB)-SAT problem. A pseudo-Boolean formula is a conjunction of pseudo-Boolean constraints. A pseudo-Boolean constraint ψ over Boolean variables $x_0, \ldots, x_{n-1}$ is an inequality of the form:

$$\sum_{i=0}^{n-1} c_i \dot{x}_i \geq c_n,$$

where $c_0, \ldots, c_n \in \mathbb{Z}$ and $\dot{x}_i \in \{0,1\}$ (corresponding to the assignment of x_i). A pseudo-Boolean constraint ψ is satisfied if and only if the sum of the coefficients c_i with $0 \leq i < n$ for which the associated variable x_i is activated, that is $x_i = 1$, is greater or equal than c_n. A pseudo-Boolean formula Ψ_{PB} is satisfied if and only if each constraint $\psi \in \Psi_{\text{PB}}$ is satisfied.

The question whether Ψ_{PB} is satisfiable is also known as the PB-SAT problem. The application of PB-SAT is related to the application of SAT. In order to transform a circuit-oriented problem into a PB-SAT problem, the circuit's logic behavior has to be modeled in PB constraints. Each signal s_j in a circuit is assigned a Boolean variable x_j. Similar to the transformation into a SAT problem [Lar92], the functionality of each gate g can be represented by a set of constraints ψ_g. In fact, each SAT constraint, i.e. a CNF clause, can be easily converted into a PB constraint. Table 7.6 shows the representation of an AND gate in PB constraints as well as in CNF. Note that a negative literal $\overline{x}_i$ is represented by the term $(1-x_i)$. The conversion of a SAT problem in CNF into a PB-SAT problem is therefore straightforward.

The PB representation $\Psi_{\mathscr{C}}$ for circuit $\mathscr{C}$ with gates $g_1, \ldots, g_k$ is given by the following formula:

$$\Psi_{\mathscr{C}} = \prod_{j=0}^{k} \psi_{g_j}$$

In practice, $\Psi_{\mathscr{C}}$ is then extended with problem-specific constraints Ψ_F which are for example needed for fault propagation and activation as described for CNF in Chap. 4. Then, the derived PB-SAT instance Ψ_{PB} which can be given to a PB-SAT solver to compute a test is as follows:

$$\Psi_{\text{PB}} = \Psi_{\mathscr{C}} \cdot \Psi_F$$

However, there are two different types of PB-SAT solvers. Solvers like Pueblo [SS06] directly support PB constraints, while solvers like MiniSat+ [ES06] translate the PB-SAT problem into a SAT instance and apply regular SAT algorithms to find a solution. Obviously, the latter type of solvers are particularly suited for problems, which can be modeled with many CNF clauses and a few pseudo-Boolean constraints [Anj09]. Since the kind of problems considered in this book are mostly CNF-based, the latter type of PBO solvers are better suited.

The PBO problem consists of a pseudo-Boolean formula Ψ_{PB} and an objective function $\mathscr{F}$. The formula $\mathscr{F}$ is to minimize a given objective function of the form:

$$\mathscr{F}(x_0,\ldots,x_{n-1}) = \sum_{i=0}^{n-1} m_i \dot{x}_i,$$

where $m_0,\ldots,m_{n-1} \in \mathbb{Z}$. Therefore, the PBO problem is to determine the solution which satisfies Ψ (solving the PB-SAT problem) and, at the same time, minimizes the given objective function $\mathscr{F}$.

In order to find the solution which minimizes the given objective function $\mathscr{F}$, a PBO solver calculates an initial solution at first (corresponding to a PB-SAT solution) which is then improved in the following until no better solution can be found. Generally, the search space of such a problem is huge and typically many iterations are needed to find the minimum solution. However, PBO solvers use efficient conflict-based learning techniques and effective heuristics (known from SAT solvers) during the search. As a result, the search space can typically be traversed very quickly, since a large part can be pruned by learned information. Therefore, PBO solvers have the potential to cope with the high complexity of the timing-aware ATPG problem.

7.3.3 PBO Formulation: Timing-Aware ATPG

This section describes how the timing-aware ATPG problem is represented as a PBO problem, i.e. as a PB-SAT instance Ψ_{PB} and a minimization function $\mathscr{F}$. We first describe how the PB-SAT instance is composed and how the minimization function is derived. Afterwards, details about the constraints which have to be added to the PB-SAT instance in order to guarantee a consistent path representation are presented.

7.3.3.1 PB-SAT and Minimization Function

The use of PB-SAT and PBO, respectively, has the advantage that the efficient solving and search space pruning technique of state-of-the-art solvers can be applied to solve the specific problem. However, the correct and complete formulation as a PBO problem instance is crucial for the efficient application. As stated above, the use of a PBO solver requires the creation of a PB-SAT instance Ψ_{PB} and a minimization function $\mathscr{F}$. The proposed PB-SAT formulation is based on the SAT formulation for ATPG proposed in TEGUS [SBS96]. As shown above, any SAT instance can be transformed into a PB-SAT instance in a straightforward manner but not vice versa. The test generation formulation consists of the following parts:

- Ψ_C describes the logic of the necessary circuit parts. Note that two consecutive time frames t_1 and t_2 have to be considered for transition test generation. A signal x is therefore associated with two variables x_1 and x_2 representing the value of the line in the corresponding time frame.
- Ψ_F describes the faulty part of the circuit. That is the fault site as well as the logic of the faulty output cone. An additional variable y^f is assigned to each signal y in the faulty output cone which represents the value of y in the faulty part.
- Ψ_D describes additional constraints necessary for fault propagation and fault observation. In particular, these constraints make sure that a D-chain exists, i.e. there exists a path from the fault site to an observation point on which the fault is propagated. An additional variable y^d (also called d-variable) is associated with each signal y in the faulty output cone. This variable is 1 if the fault is propagated to an observation point via this line.

This formulation is extended for the problem of finding the longest path through the fault site. Here, a clear path representation is needed for identifying the longest path automatically by the solver used. The last part of the formulation, i.e. Ψ_D already includes a propagation path representation by the d-variables of the output cone. When the variable y^d of signal y is assigned to 1, the fault is propagated along line y. Therefore, the propagation path is represented by the set of lines whose d-variable is 1. More formally, let Y be the set of lines in the output cone of the fault site, then the propagation path P^p is represented as follows:

$$P^p = \{y \in Y : \ y^d = 1\}$$

However, this representation has to be extended, since it covers the propagation path only. The activation path has to be considered for identifying the longest path, too. Generally, setting the desired transition value at the fault site is sufficient for the solver used to create an activation path. However, additional information is required for path identification. Therefore, a j-variable z^j is assigned to each line z in the structural support of the fault site. This is illustrated in Fig. 7.5. Note that the signal line of the fault site is assigned a d-variable as well as a j-variable. Both variables of the fault site are fixed to 1 in the problem formulation to trigger the search.

The j-variable z^j of line z is 1 if the line carries a transition along the activation path. Therefore, similar to the representation of the propagation path P^p, the activation path P^a is represented by those lines whose j-variable is assigned to 1. More formally, let Z be the set of lines in the support of the fault site, then the activation path P^a is represented as follows:

$$P^a = \{z \in Z : \ z^j = 1\}$$

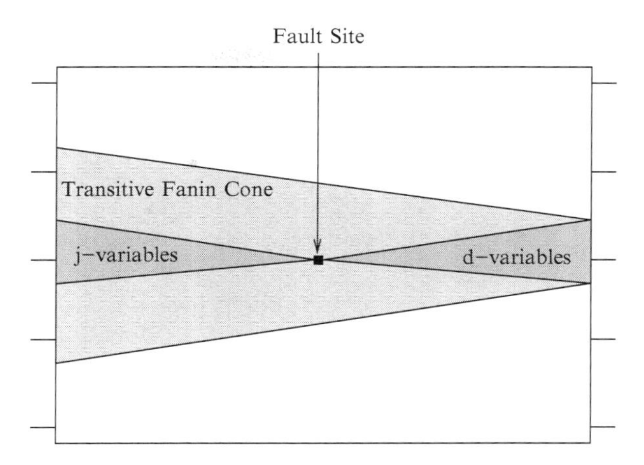

Fig. 7.5 d- and j-variables in PB-SAT transformation

Note that the constraints which guarantee the correct assignment of the d and j-variable are given below. Eventually, the complete path P^f for fault activation as well as for fault propagation is derived by the union of P^a and P^p:

$$P^f = P^a \cup P^p$$

This path representation allows the solver to identify the path by checking the assignment of the d- and j-variables. This is then used to create the minimization function which is responsible for identifying the longest path. Therefore, the minimization function $\mathscr{F}$ consists of the d- as well as of the j-variables of the given instance. In addition, to incorporate the delay aspect, each variable x in the minimization function is associated with a static delay value d^x (obtained by timing analysis) which represents the delay of the line as well as of the predecessor gate[5]:

$$\mathscr{F}(y_1^d, \ldots, y_n^d, z_1^d, \ldots, z_m^d) = \sum_{i=1}^{n} -d^{y_i} \cdot y_i^d + \sum_{j=1}^{m} -d^{z_j} \cdot z_j^j$$

The result of $\mathscr{F}$ is the accumulation of the delay values of the activated variables, i.e. those variables which are assigned to 1 in the current assignment. Given to a PBO solver, the ultimate solution is the assignment which minimizes $\mathscr{F}$. This directly corresponds to the longest path through which the transition fault is detected.

[5]Note that the delay value is given in $\mathscr{F}$ as a negative value, since state-of-the-art PBO solvers typically perform minimization but not maximization.

7.3.3.2 Constraints for Consistent Path Representation

This section shows which constraints or implications have to be added to the PB-SAT instance to guarantee a correct and consistent path representation. This includes the following properties:

- It has to be guaranteed that the transition is activated and propagated along at least one path. These constraints are needed for fault detection and are described in the following by Ψ_{path}.
- It has to be ensured that the d- and j-variables of *exactly one path* are assigned to 1, although there exist multiple paths along which the transition is propagated or activated, respectively. This is especially important since the minimization function $\mathscr{F}$ is defined over *all* d- and j-variables. The solver tries to assign as many as possible of these variables to 1. These constraints are described in the following by Ψ_{one}.
- Different arrival times of transitions at gate inputs have to be considered in order to make sure that the correct path which causes the transition at the output is identified. This is described by Ψ_{tran}.

In summary, the PB-SAT instance Ψ_{PB} which incorporates these properties is derived as follows:

$$\Psi_{\text{PB}} = \Psi_C \cdot \Psi_F \cdot \Psi_{\text{path}} \cdot \Psi_{\text{one}} \cdot \Psi_{\text{tran}}$$

These constraints are given in detail in the following.

At Least One Path – Ψ_{path}

First, in order to guarantee the propagation of the fault via at least one path from the fault site to an observation point, the following implications (which were used to accelerate the search in [SBS96]) have to be added to the instance Ψ_{PB} (denoted by Ψ_D).

$$(y^d = 1) \rightarrow y^g \neq y^f$$
$$(y^d = 1) \rightarrow (z_1^d + \ldots + z_k^d)$$

The first implication ensures that the value of the good circuit (y^g) is different from the value of the faulty circuit (y^f). The second implication guarantees that if the d-variable y^d of line y is 1, at least one d-variable of the successors of y, i.e. $z_1, \ldots, z_k$, has to be 1 in order to ensure the detection of the fault. Similar implications have to be added for the activation path as constraints:

$$(x^j = 1) \rightarrow x_1 \neq x_2$$
$$(x^j = 1) \rightarrow (w_1^j + \ldots + w_l^j)$$

Table 7.7 PB representation of implications

$(y^d = 1) \rightarrow y^g \neq y^f$	$((1-y^d)+(1-y^g)+(1-y^f) \geq 1)$
	$\cdot((1-y^d)+y^g+y^f \geq 1)$
$(y^d = 1) \rightarrow (z_1^d + \ldots + z_k^d)$	$((1-y^d)+z_1^d+\ldots+z_k^d \geq 1)$
$(x^j = 1) \rightarrow x_1 \neq x_2$	$((1-x^j)+(1-x_1)+(1-x_2) \geq 1)$
	$\cdot((1-x^j)+x_1+x_2 \geq 1)$
$(x^j = 1) \rightarrow (w_1^j + \ldots + w_l^j)$	$((1-x^j)+w_1^j+\ldots+w_l^j \geq 1)$

However, there is no good and faulty value for the fault activation. The desired transition is identified by comparing the value of the line in the initial time frame and the value in the final time frame. Therefore, the first implication ensures that the value of the first time frame (x_1) is different from the value of the second time frame (x_2), i.e. a transition occurs. The second implication guarantees a consistent activation path, since if the j-variable of a line is assigned to 1, at least one predecessor of x, i.e. $w_1, \ldots, w_l$, has to assume a transition.

Table 7.7 shows the PB representation of the described implications. The implications presented below can be similarly transferred.

Exactly One Path – Ψ_{one}

In order to ensure that d- and j-variables of exactly one path are assigned to 1, the following implications are needed for a propagation path. Let Y be the set of lines in the output cone. For each line $y \in Y$, the direct predecessors which are themselves in Y are given by $p_1, \ldots, p_m$. Note that the fault site is not part of Y. In addition, let Y^{fan} be the set of fanouts in the output cone and let $b_1, \ldots, b_k$ the branches of each fanout $y \in Y^{\text{fan}}$. Then, the following implications are included in Ψ_{one} for each branch and each gate in Y:

$$(b_i^d = 1) \rightarrow \overline{b}_1^d \cdot, \ldots, \cdot \overline{b}_{i-1}^d \cdot \overline{b}_{i+1}^d \cdot \ldots \cdot \overline{b}_k^d \text{ where } 0 < i < (k+1)$$

$$(y^d = 1) \rightarrow p_1^d + \ldots + p_m^d$$

The first implication ensures that for each fanout in the output cone, there is only one branch with a d-variable assigned to 1. Note that the implications are unidirectional. Therefore, it is still possible that the fault is propagated along multiple paths. However, only the d-variables of one path are activated. The second implication makes sure that d-variables can only be assigned to 1 if the d-variable of one predecessor is assigned to 1. This ensures that the beginning of the D-chain is the fault site. In summary, these constraints guarantee that the propagation path begins at the fault site and the d-variables of exactly one consistent path are assigned to 1.

The constraints needed for the activation path are different, since the direction is contrary. Let Z be the set of lines in the structural support. For each line $z \in Z$,

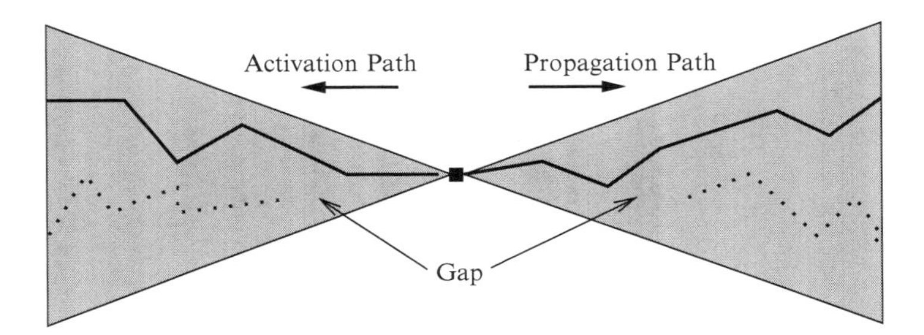

Fig. 7.6 Multiple paths

the direct predecessors are given by $p_1, \ldots, p_l$. In addition, the direct successors of z which are themselves in the support are given by $s_1, \ldots, s_k$. Then, the following implications are included in Ψ_{one}:

$$(p_i^j = 1) \rightarrow \overline{p_1^j} \cdot, \ldots, \cdot \overline{p_{i-1}^j} \cdot \overline{p_{i+1}^j} \cdot \ldots \cdot \overline{p_l^j} \text{ where } 0 < i < (l+1)$$

$$(z^j = 1) \rightarrow s_1^j + \ldots + s_k^j$$

Similar to the implications of the propagation path, the first implication ensures that only one input of each gate in the support has a j-variable assigned to 1. The second implication ensures that the end of the activation path is the fault site.

The importance of especially the second implication is illustrated in Fig. 7.6. The first implication guarantees that the path, i.e. a d and j path assignment, takes only one branch at each fanout or at each gate, respectively. The second implication excludes additional path assignments which do not have their source in the fault site as shown as pointed lines in the illustration.

Determining the Correct Transition – Ψ_{tran}

In some cases, the origin of the transition is not clear from the value assignment. This problem is demonstrated in Fig. 7.7. Here, an AND gate is shown which assumes a falling transition on the output c. However, the origin of the transition cannot be determined by the value assignment since both inputs switch in the same direction. Since the final value is the controlling value, the transition is caused by the first occurring transition. However, if the solver can freely choose the path, it would choose the longer path along input a which is wrong, since the transition is caused by input b.

Therefore, for each predecessor p_i of line y in the output cone, let $q_1, \ldots, q_n$ be those predecessors of y whose arrival time is later than the arrival time of p_i.Let further cv (ncv) be the controlling value (non-controlling value) for the predecessor gate of y. In order to exclude that the incorrect transition path is chosen, the following implication is needed:

$$((p_i)_1 = ncv) \cdot ((p_i)_2 = cv) \cdot (y^d = 1) \rightarrow (\overline{q_1}^d) \cdot \ldots \cdot (\overline{q_n}^d)$$

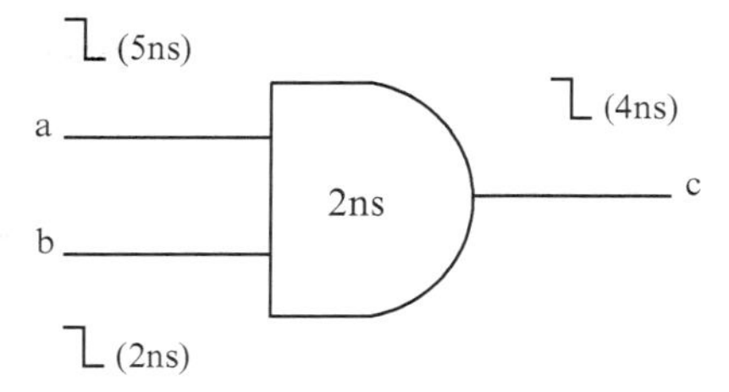

Fig. 7.7 Origin of transitions

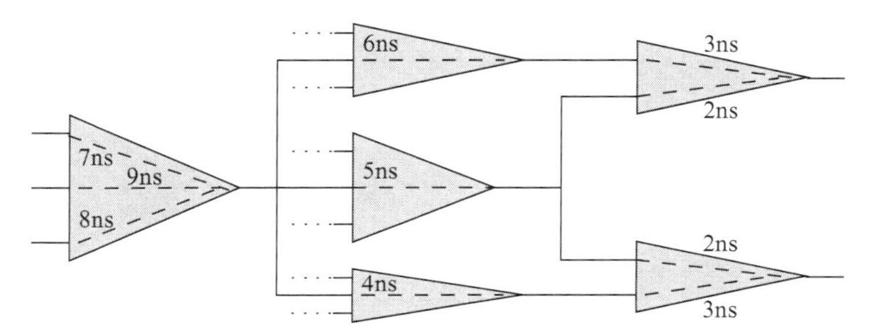

Fig. 7.8 Using FFRs

That is, if the transition on p_i goes from the non-controlling value to the controlling value and y is on a D-chain, than all other inputs of the same gate cannot be on the D-chain. This ensures that the solver has to choose the correct switching input with the smallest arrival time. Note that this does not influence the value assignment of the signal but only the decision which path is the actual D-chain. The same implication is used for the activation path with the j-variables instead of the d-variables.

7.3.4 Using Structural Information

The inclusion of the different implications in form of constraints in Ψ_{PB} increases the complexity of the PBO instance to solve. In addition, the large number of d- and j-variables increases the search space given by the minimization function. In order to reduce the search space, an improved formulation is proposed which makes use of structural information.

The circuit can be divided in *Fanout Free Regions* (FFRs). There is a unique path from every input of an FFR, i.e. a branch or primary input, to a fanout or output. Therefore, the basic idea behind the improved formulation is to restrict the d- and j-variables used in the minimization function to inputs of an FFR. The delay value which is associated with an input i of an FFR represents the delay of the path from i to the output of the FFR. This is illustrated in Fig. 7.8.

This also leads to a reduction of the additional constraints. The formulation of the constraints included in Ψ_{one} and Ψ_{tran} – which are not needed for fault detection but for path identification – can be directly applied to the FFR level instead of the gate level. This is advantageous, since the reduction of constraints typically corresponds to a run time reduction of the solving process.

7.3.5 Considering Transition-Dependent Delays

So far, only the unit delay model has been considered. In this model, a static delay value is assumed for a specific line which is independent from the value assignment or possible transition. However, the actual delay is transition dependent, e.g. the duration of a transition is typically different for a rising and a falling transition. In the following, we propose an extension of the PBO formulation for timing-aware ATPG in order to incorporate transition-dependent delays for a more realistic behavior.

The delay calculation is part of the minimization function $\mathscr{F}$ of the PBO formulation. Each d- and j-variable which is part of $\mathscr{F}$ is associated with a specific delay value. The delay values of the activated variables are then accumulated and represent the overall delay of the path. We propose to add another variable layer in order to incorporate transition-dependent delays. Each d- and j-variable is assigned two new variables d_F, d_R and j_F, j_R, respectively. These variables represent the direction of the transition: d_F, j_F (falling), d_R, j_R (rising). Then, the minimization function $\mathscr{F}$ is only defined over these variables and not over the d- and j-variables anymore.

Each of the variables d_F, d_R and j_F, j_R are then associated with the delay of the specific transition $d_{\{F,R\}}$ which is represented. Therefore, the new minimization function is formulated as follows:

$$\mathscr{F} = \sum_{i=1}^{n} -d_F^{y_i} \cdot (y_i^d)_F + -d_R^{y_i} \cdot (y_i^d)_R + \sum_{j=1}^{m} -d_F^{z_j} \cdot (z_j^j)_F + -d_R^{z_j} \cdot (z_j^j)_R$$

In order to guarantee that the new variables d_F, d_R and j_F, j_R always assume the correct value, the following implications are added to the PB-SAT instance:

$$\begin{aligned}
(y^d = 1) \cdot (y_g = 0) &\rightarrow y_F^d \\
(y^d = 0) &\rightarrow \overline{y}_F^d \\
(y^d = 1) \cdot (y_g = 1) &\rightarrow y_R^d \\
(y^d = 0) &\rightarrow \overline{y}_R^d \\
(z^d = 1) \cdot (z_g = 0) &\rightarrow z_F^d \\
(z^d = 0) &\rightarrow \overline{z}_F^d \\
(z^d = 1) \cdot (z_g = 1) &\rightarrow z_R^d \\
(z^d = 0) &\rightarrow \overline{z}_R^d
\end{aligned}$$

These implications ensure that a rising transition and an activated $d(j)$-variable always lead to an activated $d_R(j_R)$-variable. Additionally, a falling transition and an activated $d(j)$-variable always lead to an activated $d_F(j_F)$-variable. Furthermore, if the $d(j)$-variable is assigned to 0, the corresponding transition-dependent variable is assigned to 0, too.

The proposed modification of $\mathscr{F}$ leads to a more realistic delay behavior. Extending the formulation to a more fine-grained delay model is easily possible by introducing more variables and adding the corresponding implications.

7.3.6 Experimental Results

This section provides experimental results for the proposed PBO-based timing-aware ATPG approach. The approach was implemented in C++. The PBO solver used was *clasp* [GKNS07]. This solver supports both PB constraints as well as CNF constraints.[6] Therefore, clasp is also used as a SAT solver for run time comparison.

The experiments were conducted on ISCAS'89 as well as ITC'99 benchmark circuits using launch-on-capture. Industrial circuits have not been considered, since the ATPG-specific SAT techniques have not been transferred to the PBO domain yet.

All transition faults on inputs and branches were targeted. No fault dropping was performed. Table 7.8 shows the experimental results concerning run time. Column *Classic* gives the results for regular transition test generation, i.e. modeled as a SAT problem as described in Sect. 7.1, and Column *Timing-aware* gives the results of timing-aware test generation using PBO.

Column *Bits* shows the average number of specified bits of the generated tests. Column *%Bld* gives the portion of the run time needed for PBO instance generation and column *Time* shows the total run time. The overall increase in run time is given in the last column *Inc*.

Since the timing-aware ATPG problem is much more complex than regular test generation – instead of finding one arbitrary solution, the complete solution space has to be covered – the total run time is much higher up to a factor of 20. Most of the run time is spent for the solving process. The percentage of run time needed for instance generation decreases significantly at the cost of increase in solving time. For instance, half of the total run time needed for b22 was spent for building the PBO instances in regular transition test generation. Using timing-aware ATPG, the percentage decreases to only 3.4%.

Please note that the PBO solver used was applied without ATPG-specific modifications. It can be expected that – due to the similarity to the SAT problem – the integration of ATPG-specific solving techniques, e.g. similar to those proposed

---

[6]The SAT solver clasp won the gold medal in two categories of the SAT competition 2009 and in one category in the PB Evaluation 2009.

Table 7.8 Experimental results – run time

	Classic			Timing-aware			
Circ	Bits(%)	%Bld	Time(min)	Bits(%)	%Bld	Time	Inc.
s713	22	61.5	0:01	25	54.2	0:01 min	1.00x
s5378	10	64.0	0:04	10	23.3	0:07 min	1.75x
s9234	14	8.2	1:10	15	7.3	1:17 min	1.10x
s15850	4	65.0	0:39	6	3.8	7:42 min	12.01x
s35932	1	86.4	1:22	1	4.7	1:46 min	1.29x
s38417	2	70.4	1:36	2	4.0	8:22 min	5.21x
s38584	2	59.5	1:45	2	11.8	2:38 min	1.50x
b04	28	67.7	0:01	29	4.4	0:09 min	9.00x
b05	40	44.0	0:03	43	1.9	0:55 min	20.92x
b14	26	68.1	3:24	35	3.4	1:09 h	20.36x
b15	24	34.5	7:04	26	2.0	2:04 h	17.61x
b17	8	33.3	27:38	8	2.5	5:46 h	12.54x
b20	26	52.6	13:11	33	3.3	3:29 h	15.90x
b21	28	51.7	13:43	35	3.3	3:34 h	15.60x
b22	19	50.1	21:46	24	3.4	5:17 h	14.59x

in Chaps. 5 and 6, or heuristics will accelerate the search process significantly. The development of ATPG-specific PBO techniques and heuristics is therefore future work.

The advantage of the proposed method can clearly be seen by the measured path lengths of the generated test patterns. These results are presented in Table 7.9. Column *Av.P* gives the average length of the propagation path, i.e. from the fault site to an observation point while column *Av.A* states the average length of the activation path, i.e. the length from the input to the fault site. Columns *Max.P* and *Max.A* gives the maximum length of the corresponding path. Column *Inc.* gives the total path length increase of the timing-aware ATPG approach.

The results show that the length of the activation as well as the length of the propagation path can be significantly increased by the proposed method. In particular for the larger circuits, an increase in path length up to a factor of 2x is achieved. In summary, the experimental results show that the proposed PBO-based ATPG algorithm is able to cope with the complexity of timing-aware ATPG and is able to generate high-quality test patterns.

7.4 Summary

The chapter has dealt with SAT-based ATPG for the *Transition Fault Model* (TFM). The problem of the large number of unclassified faults caused by classical ATPG algorithms is even more serious for the TFM than for the *Stuck-at Fault Model* (SAFM). The classical structural ATPG algorithm available for comparison

Table 7.9 Experimental results – path length

Circ	Classic				Timing-aware				
	Av.P	Av.A	Max.P	Max.A	Av.P	Av.A	Max.P	Max.A	Inc.
s713	6.32	12.15	20	41	11.07	17.01	66	62	1.52x
s5378	7.29	5.16	23	17	11.41	5.46	26	19	1.36x
s9234	12.37	5.70	34	27	15.03	7.01	51	39	1.22x
s15850	11.78	4.80	34	21	14.75	7.17	58	47	1.32x
s35932	11.92	2.72	23	10	13.08	3.55	25	17	1.14x
s38417	11.56	4.90	30	21	13.86	6.65	41	33	1.25x
s38584	6.33	3.87	34	29	8.08	4.44	52	40	1.23x
b04	7.12	3.73	20	15	9.26	6.64	27	23	1.47x
b05	7.52	4.63	27	20	12.16	7.16	40	30	1.59x
b14	9.60	6.66	41	39	14.29	14.45	60	58	1.77x
b15	7.95	5.42	48	45	13.27	11.16	64	59	1.83x
b17	7.23	6.50	72	62	13.54	11.79	79	77	1.85x
b20	8.70	6.53	41	53	13.69	18.84	62	60	2.14x
b21	8.95	6.39	43	50	13.81	17.88	61	61	2.07x
b22	8.75	6.48	56	51	14.37	18.02	65	62	2.13x

produces over one million of unclassified faults in total. Increasing the resources of this algorithm reduces the number to roughly the half at the cost of significantly increased run time.

This chapter has shown in detail how SAT-based ATPG can be applied for *Transition Fault* (TF) test generation and how TFs can be modeled using stuck-at faults in CNF. Experiments have shown that classical SAT-based ATPG is very promising in reducing the unclassified faults but has some serious performance weaknesses. The run time behavior of SAT-based ATPG is unreasonably high for some circuits and therefore not acceptable. The performance of SAT-based ATPG is significantly improved by the application of the SAT techniques proposed in Part II of this book. Furthermore, the number of aborts can be reduced from altogether over 1 million to under 20,000 unclassified faults. The experiments have shown that the application of SAT-based ATPG provides significant advantages for test generation for industrial circuits with an increase of the fault coverage of up to 2%.

The demonstrated robustness of SAT-based ATPG can be further leveraged by considering the problem of *Small Delay Defects* (SDDs). Common ATPG techniques either do not target the detection of SDDs at all or produce significant run time overhead by using timing information during the search process. Here, incremental SAT instance generation is proposed for enhancing the quality of the generated tests. Timing information can be used during SAT instance generation in order to prioritize long propagation paths. The technique is well scalable and improves the quality of the test set at low overhead.

Additionally, a more run time intensive timing-aware ATPG approach based on pseudo-Boolean optimization has been presented. This approach is an exact method and therefore able to guarantee the fault activation and propagation via the longest path using transition-dependent delays. By this, the quality of the test set is significantly increased.

In summary, the proposed SAT techniques are shown to be able to overcome the problem of the large number of unclassified faults and clearly improve the test generation process for transition faults. Additionally, SAT and PBO formulations have been presented which are able to produce high-quality test patterns and show the potential in this domain.

Chapter 8
Path Delay Fault Model

Delay testing has become an important issue in the post production test due to the increased speed and continuously shrinking feature size of modern circuits. Several delay fault models have been proposed. Among them, the most accurate fault model is the *Path Delay Fault Model* (PDFM). This fault model captures small as well as large delay defects distributed along one path in the circuit. However, the number of paths in modern circuits is typically excessively large. Test generation for all paths is therefore not possible because of the long ATPG run time and the large number of generated tests. For that reason, usually only tests for critical paths are generated to ensure the correct timing behavior of signals on these paths. Furthermore, tests for the PDFM are used for diagnostic reasons if the timing behavior of particular paths should be verified. High-quality tests are required especially for diagnosis.

Concerning the quality, tests for PDFs can be roughly classified in two categories [KC98]: non-robust and robust. Both categories differ in the sensitization criteria of the path. Modeling static values for the robust sensitization criterion makes the robust test generation harder. Nonetheless, robust tests provide a higher quality and are therefore more desirable.

As already shown for the *Stuck-At Fault Model* (SAFM) and the *Transition Fault Model* (TFM), structural ATPG algorithms reach their limits when applied to modern designs. The number of faults which cannot be classified grows. It has been already shown in the previous chapters of this book, that SAT-based ATPG is a promising alternative to structural ATPG being fast and very robust even on large industrial circuits.

This chapter deals with the problem of generating non-robust and robust tests for *Path Delay Faults* (PDFs) in an industrial environment. Modeling only Boolean values is insufficient in such an environment. The additional values U (unknown) and Z (high impedance) have to be considered, too, as explained in Chap. 4. Furthermore, static values have to be guaranteed for robust tests. These additional values can be modeled by a set of multiple-valued logics [EFG+10]. However, the use of multiple-valued logics necessitates the application of Boolean encodings so that powerful Boolean SAT solvers can be used for high quality PDF test generation.

S. Eggersglüß and R. Drechsler, *High Quality Test Pattern Generation and Boolean Satisfiability*, DOI 10.1007/978-1-4419-9976-4_8,

Generally, the more values a multiple-valued logic has, the larger is the resulting SAT instance. A larger SAT instance typically leads to a run time increase.

Considering all industrial particularities and the robust sensitization condition results in a significant overhead for the SAT instance as well as for the ATPG run time. A SAT instance generation flow [ED12] is presented which makes use of structural properties of the circuit. This flow is able to significantly decrease the size of the SAT instance and makes SAT-based ATPG feasible for high-quality test pattern generation.

The techniques presented in this chapter were implemented as the SAT-based ATPG tool MONSOON [EFG+10]. It is shown that MONSOON outperforms a state-of-the-art PDF ATPG tool by several factors on average. Further experiments on large industrial circuits show the efficiency and robustness even for the generation of robust tests.

Besides the features of the tool MONSOON, a new quality level is introduced. Since not all paths are robustly testable, typically, a non-robust test is generated for a robustly untestable path. However, there is a large quality gap between non-robust and robust tests. This chapter shows the new quality level of *As-Robust-As-Possible* (ARAP) tests [ED11a] which diminishes the quality gap. Accompanied with the introduction of ARAP tests, a new test generation procedure based on *Pseudo-Boolean Optimization* (PBO) is presented which is able to automatically generate a test with highest quality possible in one step.

This chapter is structured as follows. At first, Sect. 8.1 briefly reviews related work concerning SAT-based ATPG approaches for the PDFM. Then, SAT-based ATPG for generating non-robust tests is described in Sect. 8.2. The modeling of static values as well as industrial particularities needed for robust test generation in industrial application is presented in Sect. 8.3. The SAT instance generation flow incorporating structural properties of the circuits is introduced in Sect. 8.4. Section 8.5 introduces ARAP tests and a corresponding PBO-based ATPG approach. Section 8.6 gives experimental results of MONSOON as well as of ARAP test generation. The summary of this chapter is given in Sect. 8.7.

8.1 Related Work

This section describes SAT-based approaches for generating tests for the PDFM. The main difference to stuck-at and transition test generation is that no faulty circuit part has to be modeled but different quality levels have to be accounted for. Modeling the circuit in CNF and setting certain values in order to force the desired path sensitization is sufficient for generating a PDF test. At first, an approach for generating robust tests for combinational circuits only is briefly presented. Then, the use of incremental SAT and learning possibilities is described. Finally, ATPG approaches using the circuit SAT solver CSAT are introduced.

Table 8.1 Encoding of $\mathscr{L}_7$ [CG96]

$\eta_{\mathscr{L}_7}$	$s0$	$s1$	$\overline{s}0$	$\overline{s}1$	$x0$	$x1$	xx
s	1	1	0	0	x	x	x
v	0	1	0	1	0	1	x

Table 8.2 CNF representation of an AND gate using $\eta_{\mathscr{L}_7}$ [CG96]

$$(\overline{a}^s + \overline{b}^s + c^s) \cdot (\overline{a}^s + a^v + c^s) \cdot (\overline{b}^s + b^v + c^s) \cdot$$
$$(a^s + b^s + \overline{c}^s) \cdot (a^s + \overline{b}^v + \overline{c}^s) \cdot (\overline{a}^v + b^s + \overline{c}^s) \cdot$$
$$(\overline{a}^v + \overline{b}^v + c^v) \cdot (a^v + \overline{c}^v) \cdot \quad (b^v + \overline{c}^v)$$

8.1.1 Robust Tests in Combinational Circuits

The first approach was presented in [CG96]. Here, a SAT formulation for robust test generation in combinational circuits was proposed. Static values have to be modeled for robust tests as described in Sect. 2.3.2. Structural algorithms typically use a multiple-valued logic to model such values. In this approach, the seven-valued logic

$$\mathscr{L}_7 = \{s0, \overline{s}0, s1, \overline{s}1, x0, x1, xx\}$$

and an encoding based on two ternary variables s, v are applied to model static values in CNF. The encoding used is shown in Table 8.1. The application of an encoding leads to a larger CNF representation. The CNF representation of an AND gate $a \cdot b = c$ using $\mathscr{L}_7$ and the encoding presented in Table 8.1 is given in Table 8.2. However, the logic used is only suitable for combinational circuits. The behavior of sequential circuits cannot be modeled adequately, because the value of the first time frame is not explicitly encoded.

8.1.2 Incremental SAT and Learning

The approach presented in [KWMS00] introduces the use of *Incremental SAT* (ISAT) (see also Sect. 3.3.4) for path sensitization problems. Here, the path is sensitized incrementally and the search benefits from the previously solved SAT instances. If an unsensitizable partial path p is found, all paths containing p are also unsensitizable and do not have to be checked.

The work presented in [CH05] extends the idea and proposes the learning of unsensitizable path segments. If an unsensitizable path is found, an unsatisfiable core is generated from the unsatisfiable SAT instance and by this, an unsensitizable path segment is identified. These path segments are learned and applied in the further test generation. Paths containing such an unsensitizable path segment are unsensitizable as well. Both approaches benefit significantly from the fact that there is a large number of similar paths in the circuit [FPR94]. However, in modern circuits, the number of paths is too large for generating tests for all of them. Usually, a subset of critical paths is chosen. If only a subset of paths is chosen for test generation, the presented approaches can typically not exploit their full advantage.

8.1.3 PDF Test Generation Using CSAT

The tool KF-ATPG was proposed in [YCW04]. This tool utilizes the circuit-based SAT solver CSAT [LWCH03, LWC+03] (see Sect. 3.5) to classify PDFs under the non-robust sensitization model. The tool is applied in order to sensitize the longest propagation path from a transition fault site to an output and by this, to increase the test quality of the test set. A false-path pruning technique is used to record partial false paths (unsensitizable path segments). This pruning technique is based on a stepwise path-tracing technique similar to the technique proposed in the structural PDF ATPG DYNAMITE [FFS91]. If a conflict is detected during the path-tracing procedure, the prime unsensitizable path segment is identified by manipulating the current path and checking whether the manipulated path is still unsensitizable. The unsensitizable segments are recorded in a circuit graph. As a result, many propagation paths can be pruned early.

The work presented in [LHL07] evaluates the relationship between the robust, non-robust and functional sensitization criterion for PDF test generation using CSAT. Path delay test generation typically requires a pre-selection of the sensitization criterion. If this fails, i.e. the PDF is untestable under this sensitization criterion, a different sensitization model has to be selected and the ATPG process has to be started anew. This is a time-consuming task. Therefore, in [LHL07], a unified sensitization model is proposed that avoids to rebuild the ATPG instance for each sensitization. However, the underlying SAT engine CSAT does not support static values. Therefore, robust test patterns cannot be guaranteed using this approach.

8.2 Non-Robust Test Pattern Generation

This section presents the SAT formulation for generating non-robust tests for the PDFM. Two time frames are needed for a delay test as already described in Chap. 7. Therefore, two Boolean variables x_s^1, x_s^2 are assigned to each connection $s \in \mathscr{C}$; each describing the value of s in the corresponding time frame. Then, an *Iterative Logic Array* (ILA) is formed (see Sect. 7.1). Note that the procedure is similar to the procedure applied for *Transition Fault* (TF) test generation presented in the previous chapter. The CNF for each gate is duplicated using the respective variables resulting in the CNF $\Phi_{\mathscr{C}_1}$ for the initial time frame and $\Phi_{\mathscr{C}_2}$ for the final time frame. Additional constraints Φ_{seq} describe the functionality of the flip-flops in order to guarantee the correct sequential behavior. These constraints ensure the equivalence of the value of a pseudo primary output in t_1 and the value of the corresponding pseudo primary input in t_2. This results in the following formula for the unrolled circuit $\mathscr{C}_t$:

$$\Phi_{\mathscr{C}_t} = \Phi_{\mathscr{C}_1} \cdot \Phi_{\mathscr{C}_2} \cdot \Phi_{\text{seq}}$$

Fault specific constraints have to be added to $\Phi_{\mathscr{C}_t}$ in order to generate a test for a PDF $(\mathscr{P}, \{\uparrow, \downarrow\})$ with $\mathscr{P} = \{g_1, \ldots, g_k\}$. The fault specific constraints can be

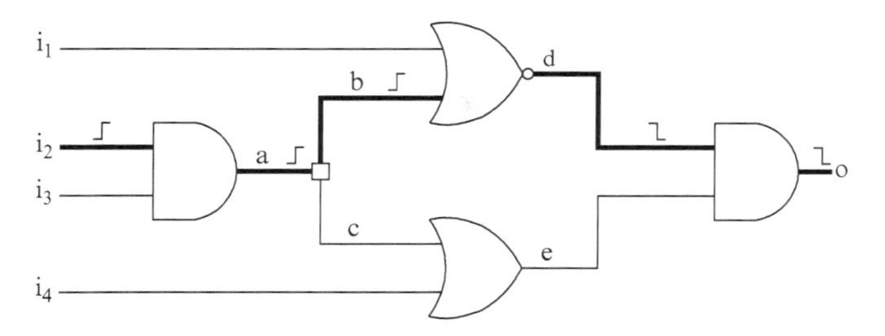

Fig. 8.1 Example circuit $\mathscr{C}$ for non-robust test generation

considered as fixed assignments to variables and are divided into two parts. The transition must be launched at g_1 (Φ_{tran}) and the side inputs of $\mathscr{P}$ have to be assigned according to the non-robust sensitization criterion as given in Table 2.2 (denoted by Φ_o). Finally, the non-robust test generation problem is formulated as follows:

$$\Phi_{(\mathscr{P},\{\uparrow,\downarrow\})} = \Phi_{\mathscr{C}_t} \cdot \Phi_{\text{tran}} \cdot \Phi_o$$

If $\Phi_{(\mathscr{P},\{\uparrow,\downarrow\})}$ is satisfiable, $\mathscr{P}$ is a non-robustly testable path and the test for fault $(\mathscr{P},\{\uparrow,\downarrow\})$ can be created directly from the calculated solution by extracting the assignments of the input variables.

Example 8.1. Consider the example circuit $\mathscr{C}$ shown in Fig. 8.1 which is a replication of the circuit given in Fig. 2.8. The path under test is $\mathscr{P} = \{i_2, a, b, d, o\}$ and the PDF for which test generation is performed is described by $(\mathscr{P},\uparrow)$. The CNF of the circuit is as follows[1]:

$$\Phi_{\mathscr{C}_1} = \Phi_a^{\text{AND}_{t1}} \cdot \Phi_d^{\text{NOR}_{t1}} \cdot \Phi_e^{\text{OR}_{t1}} \cdot \Phi_o^{\text{AND}_{t1}}$$
$$\Phi_{\mathscr{C}_2} = \Phi_a^{\text{AND}_{t2}} \cdot \Phi_d^{\text{NOR}_{t2}} \cdot \Phi_e^{\text{OR}_{t2}} \cdot \Phi_o^{\text{AND}_{t2}}$$

Because no flip-flops are contained in this circuit, the equation $\Phi_{\text{seq}} = 1$ holds. The fault specific constraints for the rising transition are:

$$\Phi_{\text{tran}} = \left(\overline{i_2^1}\right) \cdot \left(i_2^2\right)$$
$$\Phi_o = \left(\overline{i_1^2}\right) \cdot \left(i_3^2\right) \cdot \left(e^2\right)$$

A corresponding test given by the solution of the SAT solver for $\Phi_{(\mathscr{P},\uparrow)}$ is:

$$V_1 = \{i_1^1 = 0, i_2^1 = 0, i_3^1 = 0, i_4^1 = 1\}$$
$$V_2 = \{i_1^2 = 0, i_2^2 = 1, i_3^2 = 1, i_4^2 = 0\}$$

[1] Note that both fanout branches b and c can be described by the variable a for reasons of simplicity.

The application of SAT-based algorithms to non-robust test generation is straightforward, because the problem can be directly modeled in Boolean logic. However, ATPG for robust tests is more complex, since non-Boolean values have to be considered. SAT-based ATPG for robust test generation is presented next.

8.3 Robust Test Pattern Generation

According to the robust sensitization criterion given in Table 2.2, considering Boolean values only is not sufficient for modeling static values. Two discrete points in time t_1, t_2 are modeled using Boolean values, but no information about the transitions between t_1 and t_2 is given. Therefore, an ILA representation as used for non-robust PDF test generation and TF test generation (see Chap. 7) is not directly applicable. The following example motivates the use of a multiple-valued logic.

Example 8.2. Consider the AND gate in Fig. 8.2. If the robust sensitization criterion requires that the output is set to a static 0 value ($S0$), setting both output variables corresponding to the two time frames to 0 is insufficient.[2] Then, a rising and a falling transition on the inputs would satisfy the condition, because the controlling value is assumed in t_1, t_2 on different inputs as shown in Fig. 8.2a. However, if the inputs do not switch simultaneously, which cannot be guaranteed without timing information, a glitch could be produced on the output.

This case must be excluded by explicitly modeling static values. This ensures that a static value on the output of a gate has its source in one or more static values on the inputs. This is shown at the AND gate in Fig. 8.2b.

Static values can be handled using the multiple-valued logic $\mathscr{L}_{6s}$:

$$\mathscr{L}_{6s} = \{S0, 00, 01, 10, 11, S1\}$$

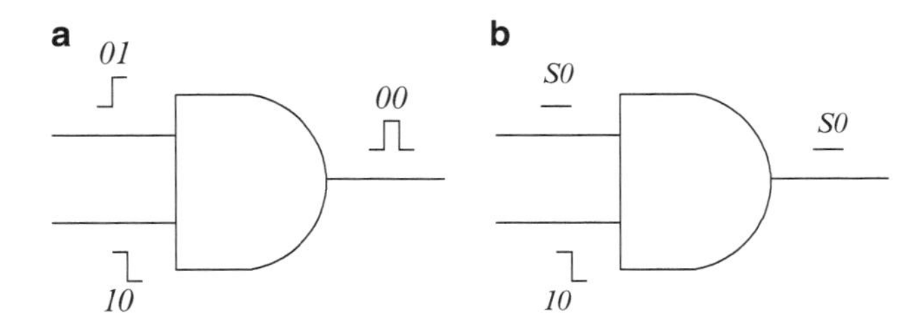

Fig. 8.2 Guaranteeing static values. (**a**) Boolean modeling and (**b**) explicit static values

[2]This simplification is done in [LHL07].

Table 8.3 Truth table for an AND gate in $\mathscr{L}_{6s}$

AND	$S0$	00	01	10	11	$S1$
$S0$	$S0$	$S0$	$S0$	$S0$	$S0$	$S0$
00	$S0$	00	00	00	00	00
01	$S0$	00	01	00	01	01
10	$S0$	00	00	10	10	10
11	$S0$	00	01	10	11	11
$S1$	$S0$	00	01	10	11	$S1$

The name of the value determines the signal's behavior in t_1 and t_2. The first position gives the value of the connection in t_1, whereas the second position describes the value in t_2. The values $S0$ and $S1$ represent the static values. The truth table for an AND gate modeled in $\mathscr{L}_{6s}$ is presented in Table 8.3.

However, modeling the circuit in $\mathscr{L}_{6s}$ is not sufficient for an industrial application. As already described in Sect. 4.2, unknown values and tri-state elements have to be considered for test generation. Chapter 4 presented a four-valued logic $\mathscr{L}_4 = \{0, 1, U, Z\}$ used for modeling stuck-at faults in industrial circuits. Each connection in both time frames is modeled in $\mathscr{L}_4$. However, $\mathscr{L}_4$ is only suitable for modeling one time frame. The Cartesian product of all values in $\mathscr{L}_4$ is needed to represent the behavior of the signal in two time frames. This results in the following 16-valued logic $\mathscr{L}_{16}$:

$$\mathscr{L}_{16} = \{00, 01, 10, 11, 0U, 1U, U0, U1, UU, \\ 0Z, 1Z, Z0, Z1, UZ, ZU, ZZ\}$$

However, $\mathscr{L}_{16}$ can only be used for non-robust test generation in industrial application. For robust test generation, $\mathscr{L}_{16}$ has to be extended with the static values $S0, S1$ and SZ resulting in the 19-valued logic $\mathscr{L}_{19s}$:

$$\mathscr{L}_{19s} = \{S0, 00, 01, 10, 11, S1, 0U, 1U, U0, U1, UU, \\ 0Z, 1Z, Z0, Z1, UZ, ZU, ZZ, SZ\}$$

Similar to the use of the four-valued logic $\mathscr{L}_4$, a Boolean encoding has to be used to transform the multiple-valued problem into a Boolean problem. However, five Boolean variables x^1, x^2, x^3, x^4, x^5 are needed to represent all possible states of the signal. The encoding $\eta_{\mathscr{L}_{19s}}$ used in this book is shown in Table 8.4.[3]

The precise SAT formulation for robust test generation is similar to the formulation for non-robust test generation with the difference that the CNF circuit representation has to be used which is able to model static values. Furthermore, the robust sensitization condition has to be assumed at the side inputs of $\mathscr{P}$.

[3]The procedure which is presented in [ED08] was applied to identify efficient Boolean encodings. The encoding used yielded the best results in the experimental evaluation.

Table 8.4 Boolean encoding $\eta_{\mathscr{L}_{19s}}$ for $\mathscr{L}_{19s}$

var	$S0$	00	01	10	11	$S1$	0U	1U	U0	U1	UU	0Z	1Z	Z0	Z1	UZ	ZU	ZZ	SZ
x^1	0	1	0	1	0	0	1	1	1	0	1	0	1	0	1	1	0	0	1
x^2	0	1	1	1	1	0	0	0	1	0	0	1	0	0	0	1	0	0	1
x^3	0	0	0	1	1	1	0	1	1	1	1	1	1	1	0	1	0	0	0
x^4	1	1	1	1	1	1	1	1	0	0	0	0	0	0	0	1	1	0	0
x^5	0	0	0	0	0	0	0	0	0	0	0	1	1	1	1	1	1	1	1

8.4 SAT Instance Generation Flow

In principle, $\mathscr{L}_{19s}$ can be used to model the circuit for SAT-based ATPG. However, logics with less values are generally more compact in their CNF representation than logics with more values. In fact, the exclusive use of $\mathscr{L}_{19s}$ would result in excessively large SAT instances and typically in run times too large for practical application. For solving this problem, the concept of logic classes and the usage of several multiple-valued logics is introduced.[4]

Typically, only a few connections in a circuit can assume all values contained in $\mathscr{L}_{19s}$ or $\mathscr{L}_4$. For example, there are only very few gates that are able to assume the value Z. Modeling all gates with $\mathscr{L}_{19s}$ or $\mathscr{L}_4$ would be correct but would also unnecessarily blow up the CNF representation. This is because the majority of all gates are not able to assume all possible values. These elements can be modeled in a multiple-valued logic with a smaller number of values. Therefore, a structural analysis is applied as a pre-processing step to determine for each gate which values can be assumed [EFG+10]. In this step, sources of U and Z values are identified and the values are propagated through the circuit to mark those elements which can potentially assume these values. This step has to be done only once and the results can be reused for each fault model and each quality level. Four different logic classes are identified for the classification of gates:

$$\mathrm{LC}_Z, \mathrm{LC}_{U1}, \mathrm{LC}_{U2}, \mathrm{LC}_B$$

- LC_Z – A gate g belongs to LC_Z if all values can be assumed in t_1 and t_2. Only tri-state elements belong to this class.
- LC_{U1} – A gate g belongs to LC_{U1} if the values $0, 1$ and U can be assumed in t_1 and t_2, but not Z.
- LC_{U2} – A gate g belongs to LC_{U2} if the values 0 and 1 can be assumed in t_1 and t_2, whereas U can be assumed only in t_2. A gate is in LC_{U2} if a flip-flop in the fanin cone of the gate is no source of unknowns but can be fed by an unknown from t_1. As a result, the value U can be propagated only in t_2.
- LC_B – A gate g belongs to LC_B if only 0 and 1 can be assumed in t_1 and t_2.

[4] This concept is a generalization of the concept of the hybrid logic presented in Sect. 4.2.2.

Table 8.5 Multiple-valued logics and logic mapping

Logic	Value set	Stuck-at	Transition	Path delay non-rob.	Path delay robust
$\mathscr{L}_{19s}$	$\mathscr{L}_{16} \cup \{S0,S1,SZ\}$	–	–	–	LC_Z
$\mathscr{L}_{16}$	$\mathscr{L}_{9} \cup \{0Z,1Z,Z0,Z1,UZ,ZU,ZZ\}$	–	LC_Z	LC_Z	–
$\mathscr{L}_{11s}$	$\mathscr{L}_{9} \cup \{S0,S1\}$	–	–	–	LC_{U1}
$\mathscr{L}_{9}$	$\mathscr{L}_{6} \cup \{U0,U1,UU\}$	–	LC_{U1}	LC_{U1}	
$\mathscr{L}_{8s}$	$\mathscr{L}_{6} \cup \{S0,S1\}$	–	–	–	LC_{U2}
$\mathscr{L}_{6}$	$\mathscr{L}_{4B} \cup \{0U,1U\}$	–	LC_{U2}	LC_{U2}	
$\mathscr{L}_{6s}$	$\mathscr{L}_{4B} \cup \{S0,S1\}$	–	–	–	LC_B
$\mathscr{L}_{4B}$	$\{00,01,10,11\}$	–	LC_B	LC_B	–
$\mathscr{L}_{4}$	$\{0,1,U,Z\}$	LC_{U1}, LC_Z	–	–	–
$\mathscr{L}_{B}$	$\{0,1\}$	LC_B, LC_{U2}	–	–	–

During circuit-to-CNF transformation, the logic class is mapped to a multiple-valued logic according to the fault model used and the desired quality level. Table 8.5 shows a summary of the logics used and the corresponding value sets. The table further shows the logic mapping which is used to create a CNF representation as compact as possible. Besides the PDFM which is the main focus of this chapter, the table also includes the logic mapping for stuck-at and transition fault test generation, since this flow can also be used for these fault models. For instance, $\mathscr{L}_B$ is used for stuck-at faults for all gates in LC_{U2}. In contrast, if robust path delay tests are desired, $\mathscr{L}_{8s}$ has to be used for gates in LC_{U2}. The application to other fault models as well shows the flexibility of the given flow.

The complete flow is illustrated in Fig. 8.3. First, the circuit is analyzed and the structural classification is done for each element. Then, according to the fault model and the desired quality, this information is used to map each logic class on a specific logic which is used for the circuit-to-CNF transformation. This procedure results in SAT instances with significant decreased size well suited for high-quality test generation for practical circuits.

8.5 As-Robust-As-Possible Test Generation

The quality of a delay test for path $\mathscr{P}$ is determined by the sensitization criterion applied to the side inputs of $\mathscr{P}$. However, the stringent conditions on the side inputs must hold for each gate $g \in \mathscr{P}$ in order to generate a robust test. Crucial is that all side inputs $s \in S_S^{\mathscr{P}}$ have to assume a static value in order to avoid that a test is invalidated by another delay fault. If at least one side input is not able to assume the required value, a robust test is not possible anymore. Then, a non-robust test has to be generated.

Typically, robust test generation fails because a subset of the side inputs (not all) cannot be set to static values at the same time because of logic correlations. For a

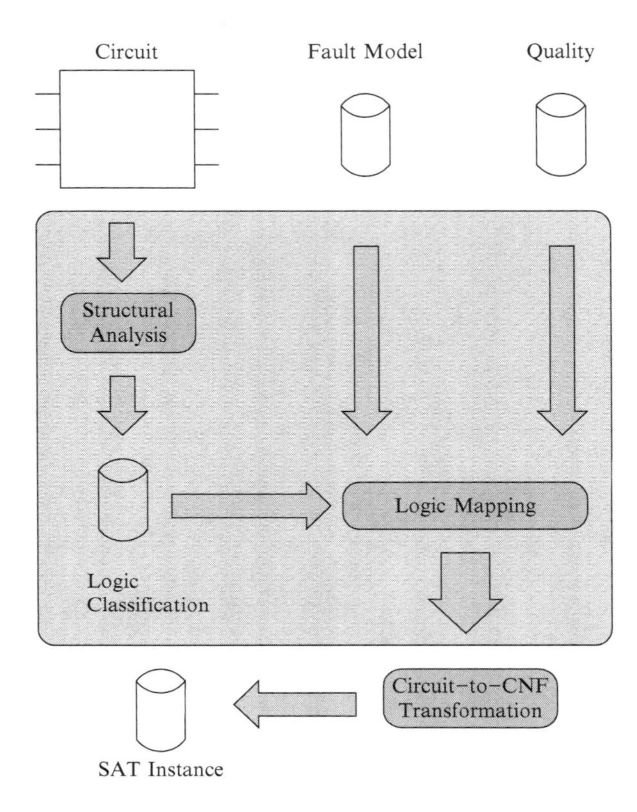

Fig. 8.3 SAT instance generation flow

non-robust test, the conditions on the side inputs are significantly relaxed for *all* side inputs of $\mathscr{P}$. Here, none of the side inputs of $\mathscr{P}$ requires a static value. As a result, an invalidation of the test is possible at any side input $s \in S_S^{\mathscr{P}}$. Therefore, there is a large quality gap between non-robust and robust tests.

Instead of generating a non-robust test, we propose to generate an *As-Robust-As-Possible* (ARAP) test for a robustly untestable fault. In contrast to a non-robust test, the number of side inputs $s \in S_S^{\mathscr{P}}$ which can be set to a static value is maximized. By this, the quality of the test is increased, since each side input set to a static value cannot be invalidated by other delay faults anymore.

Example 8.3. Figure 8.4 shows an example circuit (including one flip-flop e) with tests for a PDF with falling transition on path $a-d-g-j-m$. This PDF is robustly untestable due to the sequential behavior. A non-robust test requires that the side inputs of the gates d, g and m are set to the non-controlling value of the gate in t_2, i.e. line b and g are set to $X1$, while line l assumes the value $X0$. This is shown in Fig. 8.4a.

Although the PDF is robustly untestable, some of the inputs can be set to a static value. Figure 8.4b shows an ARAP test in which the number of static side inputs is maximized. Here, two of all three side inputs, i.e. b and c, are set to a static value

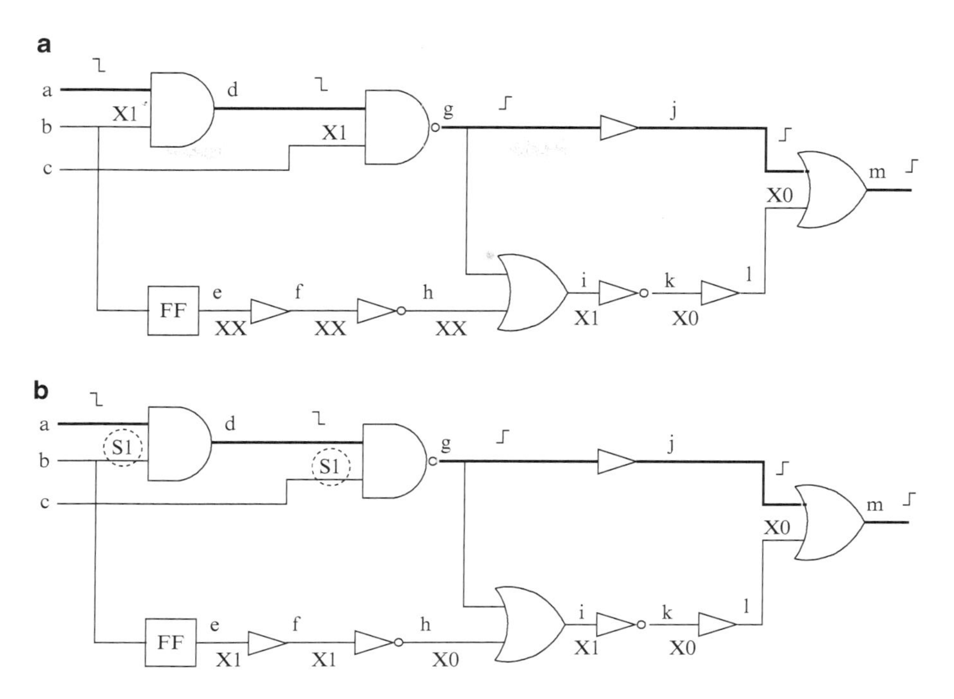

Fig. 8.4 Tests for path $a - d - g - j - m$. (**a**) Non-robust test; (**b**) ARAP test

and prevent that other delay faults invalidate the test at the corresponding gates. A static value at side input l is not possible, because of the assignment of side input b and the sequential behavior (flip-flop e).

Since neither the quality level of the test, i.e. robust or non-robust, nor – in case of a robustly untestable fault – the conflicting side inputs are known beforehand, an automated procedure is needed to efficiently generated a test with the highest quality possible. That is either a robust test or a test with a maximized number of static side inputs.

This section shows how such ARAP tests can be generated using *Pseudo-Boolean Optimization* (PBO). For an introduction to PBO, it is referred to Sect. 7.3.2. The precise PBO formulation for ARAP test generation is presented in Sect. 8.5.1. Then, Sect. 8.5.2 shows an incremental SAT formulation [EFG$^+$10] which is able to distinguish between constraints for non-robust and robust tests in an incremental manner. Section 8.5.3 shows how the presence of small delay defects can be considered in this formulation in order to decrease the possibility that the test will be invalidated by these kind of defects.

8.5.1 Test Generation for ARAP Tests

Robust tests promises high quality for PDFs. Unfortunately, the majority of all paths in a circuit is typically robustly untestable. Non-robust tests have to be generated for those paths (if one exists). As described above, there is a large quality gap between non-robust and robust tests. Therefore, the generation of ARAP tests is proposed to diminish this gap. However, the ARAP test generation problem cannot be formulated as a single SAT problem, since the maximization of static side inputs is required.

The following procedure is presented to efficiently generate ARAP tests. In order to generate an ARAP test with a maximized number of static side inputs, we formulate the problem as a *Pseudo-Boolean Optimization* (PBO) problem and solve it by a dedicated PBO solver. For a circuit $\mathscr{C}$ and a PDF F on path $\mathscr{P}$, the problem is formulated as follows: First, the circuit $\mathscr{C}$ is transformed into PB constraints $\Psi_{\mathscr{C}}$ as described above. Here, the CNF representation for non-robust test generation is used. Note that the SAT instance generation flow presented above can be used. The procedure works for Boolean circuits as well as for circuits containing multiple-valued logic, e.g. $\mathscr{L}_{16}$.

Then, $\Psi_{\mathscr{C}}$ is extended with PB constraints Ψ_{NR} for non-robust sensitization along the side inputs $S^{\mathscr{P}}$. This results in a PB-SAT instance

$$\Psi_{\mathscr{C}}^{F} = \Psi_{\mathscr{C}} \cdot \Psi_{\text{NR}}^{\mathscr{P}}$$

suitable for non-robust test generation for fault F on path $\mathscr{P}$. Note that this can still be modeled as a Boolean SAT instance. Since the aim is to generate ARAP tests, the robust sensitization has to be modeled, too. Instead of using a multiple-valued logic containing additional static values, i.e. $S0, S1$ and SZ, static values are modeled separately.

Additional to the values needed for the circuit representation, one variable x^S is assigned to each signal $x \in \mathscr{C}$. The assignment of the variable x^S denotes whether the signal is guaranteed to be static. The constraints Ψ_{SVJ} (*Static Value Justification*) which are needed to determine the correct value of these variables are presented in detail in Sect. 8.5.2. Distinguishing the constraints for non-robust and robust test generation allows for an incorporation of the robust sensitization condition into the optimization function which is described next.

The set of side inputs $S_S^{\mathscr{P}}$ which have to assume a static value for a robust test is identified and the objective function $\mathscr{F}$ is formulated. The objective function $\mathscr{F}$ is formulated over the variables of the side inputs $S_S^{\mathscr{P}}$. However, not all variables of the signals contained in $S_S^{\mathscr{P}}$ have to be included in $\mathscr{F}$. For a signal x, it is sufficient to include the variable x^S in $\mathscr{F}$, since the assignment of x^S determines whether the signal is guaranteed to be static. The number of static side inputs has to be maximized. Therefore, the objective function $\mathscr{F}$ is the sum of all variables x^S of all signals $s_i \in S_S^{\mathscr{P}}$ multiplied by the constant (-1), i.e.

$$\mathscr{F}(x_0^S, \ldots, x_n^S) = \sum_{i=0}^{n} (-1) \cdot x_i^S,$$

where x_i^S is the variable x^S of signal x_i and $n = |S_S^{\mathscr{P}}|$. As a result, the constants associated with each variable x_i^S of the set of side inputs are accumulated if x_i^S is active, i.e. $x_i^S = 1$, which means that the corresponding side input is guaranteed to be static.

Finally, the resulting PBO problem, i.e. the PB-SAT instance

$$\Psi_{\text{test}}^{F} = \Psi_{\mathscr{C}} \cdot \Psi_{\text{SVJ}} \cdot \Psi_{\text{NR}}^{\mathscr{P}}$$

and the objective function $\mathscr{F}$, is given to a PBO solver in order to calculate a solution. The solution space of $\Psi_{\text{test}}^{\mathscr{P}}$ covers all the solutions, i.e. non-robust as well as robust tests, for fault F on $\mathscr{P}$. The task of the objective function $\mathscr{F}$ is to pick the best solution out of the solution space. That is, if the test is robustly testable, a robust test is generated. Otherwise, an ARAP test with a maximized number of static side inputs is generated or the fault is proven untestable. The advantage of this formulation is that it has not to be distinguished between non-robust and robust test generation anymore. Both test problems are integrated into one problem instance and the generated test is automatically of the highest quality possible. The following example demonstrates the procedure.

Example 8.4. Again, consider the circuit $\mathscr{C}$ shown in Fig. 8.4 and the given falling PDF on path $a-d-g-j-m$. Each signal line x in $\mathscr{C}$ is represented by three Boolean variables x_1, x_2 and x^S. Here, x_1 (x_2) represents the value of x in the initial (final) time frame. The assignment of the variable x^S determines whether the value at x is guaranteed to be static. In order to generate an ARAP test, the circuit is transformed into PB constraints gate by gate:

$$\Psi_{\mathscr{C}} = \psi_d \cdot \psi_e \cdot \psi_f \cdot \psi_g \cdot \psi_h \cdot \psi_i \cdot \psi_j \cdot \psi_k \cdot \psi_l \cdot \psi_m$$

The constraints for generating a non-robust test for the falling PDF on path $a-d-g-j-m$ are added (b = X1,c = X1,l = X0):

$$\Psi_{\text{NR}}^{\mathscr{P}} = (b \geq 1) \cdot (c \geq 1) \cdot ((1-l) \geq 1)$$

Finally, the objective function is formulated:

$$\mathscr{F}(b^S, c^S, l^S) = (-1) \cdot b^S + (-1) \cdot c^S + (-1) \cdot l^S$$

A solution for $\Psi_{\mathscr{C}} \cdot \Psi_{\text{SVJ}} \cdot \Psi_{\text{NR}}^{\mathscr{P}}$ with respect to $\mathscr{F}$ determined by a PBO solver provides the ARAP test with maximized number of static side inputs:

$$\text{Test} = \{a = 10, b = S1, c = S1, e = X1\}$$

8.5.2 Incremental SAT Formulation for Static Value Justification

In order to make use of the optimization function of the PBO problem, the non-robust and robust sensitization conditions have to be distinguished. This has been described in the previous section. The PB-SAT instance is formulated such that the solution space covers all non-robust tests – and by this also all robust tests. In order to determine the optimal solution, i.e. the solution with a maximum number of static side inputs, one additional variable x^S is assigned to each signal line x. The meaning of this variable is to detect whether this line is guaranteed to be static.

Additional constraints Ψ_{SVJ} for static value justification have to be added incrementally to the PB-SAT instance for non-robust test generation. These constraints add the functionality to guarantee static values and, consequently, the possibility to generate robust tests. In contrast to the multiple-valued logic presented in Sect. 8.3, these incremental constraints allow for an identification of a static value by checking the assignment of one variable, i.e. by checking x^S, and by this, the incorporation into the objective function $\mathscr{F}$.

The use of incremental SAT is not new in the field of ATPG for PDFs. However, in previous works, e.g. [KWMS00], path segments are added incrementally to speed up test generation for non-robust tests considering all paths or a large number of overlapping paths. The idea of incrementally adding constraints for robust test generation was introduced in [EFG$^+$10]. However, it was used to speed up the solving process of robust test generation by reusing learned information from non-robust test generation. The approach proposed here uses a similar incremental SAT formulation but is used for increasing the quality of the test. Contrary to the approach in [EFG$^+$10], the test is also generated in one step instead of two steps.

In the following, a description how to derive Ψ_{SVJ} is given. The additional variable x^S determines whether the signal on the connection is static ($x^S = 1$). If a static signal has to be forced on a side input g, x_g^S is fixed to 1. Additional implications are added for each gate in $\mathscr{F}(g)$ in order to justify this value. These implications are as follows for a gate g with direct predecessors $p_1, \ldots, p_n$. If the non-controlling value ncv is on the output of g, all direct predecessors $p_1, \ldots, p_n$ have to be statically non-controlling between both time frames, too:

$$\left(x_g^S = 1 \wedge g = ncv\right) \rightarrow \prod_{i=1}^{n} x_{p_i}^S = 1 \tag{8.1}$$

If the controlling value cv is on the output of g, at least one predecessor p_i has to be statically controlling:

$$\left(x_g^S = 1 \wedge g = cv\right) \rightarrow \sum_{i=1}^{n} \left(\left(x_{p_i}^S = 1\right) \cdot \left(p_i = cv\right)\right) \tag{8.2}$$

Table 8.6 CNF size of incremental SAT formulation

Logic class	$\eta_{\mathscr{L}_{16}}$		Ψ_{SVJ}		$\eta_{\mathscr{L}_{16}} + \Psi_{\mathrm{SVJ}}$			$\eta_{\mathscr{L}_{19s}}$		
	#cls	#lit	#cls	#lit	#v	#cls	#lit	#v	#cls	#lit
AND										
LC_{U1}	20	50	18	58	5	38	108	4	30	97
LC_{U2}	14	35	16	52	4	30	87	3	21	71
LC_B	6	14	9	31	3	15	45	3	15	40
Busdriver										
LC_Z	86	426	25	81	5	111	507	5	114	561

Table 8.7 Boolean encoding $\eta_{\mathscr{L}_{16}}$ for $\mathscr{L}_{16}$

Var	00	01	10	11	0U	1U	U0	U1	UU	0Z	1Z	Z0	Z1	UZ	ZU	ZZ
x_1	0	0	1	1	0	1	0	0	1	1	0	0	0	1	1	1
x_2	0	1	0	1	0	0	0	1	1	1	0	1	1	0	0	1
x_3	0	0	0	0	1	1	0	1	1	1	1	0	1	1	0	0
x_4	0	0	0	0	1	1	1	1	1	0	0	1	0	0	1	1

Thus, the justification of a static value on a side input is guaranteed. Since this is done for each gate in the transitive fanin cone, the static values have its origin in the inputs. By this, static values can be backtraced towards the inputs.

These additional implications are transformed into CNF or PB constraints according to the logic used and corresponding encoding of g. In the following, the constraints are given in CNF for a better readability. However, they can easily be transformed into PB constraints as described above. The total size of the constraints using incremental SAT is generally larger than using a multiple-valued logic containing static values, e.g. $\eta_{\mathscr{L}_{19s}}$, since the CNF representation is optimized.

Table 8.6 shows the effect on the size of the formula for an AND gate and a busdriver. Column $\eta_{\mathscr{L}_{16}}$ gives the size of the formula for non-robust test generation (using the Boolean encoding shown in Table 8.7) and logics derived from $\eta_{\mathscr{L}_{16}}$. Column Ψ_{SVJ} presents the size of the formula for the additional implications for static value justification. The total sizes of the formula for robust test generation with incremental SAT is presented in column $\eta_{\mathscr{L}_{16}} + \Psi_{\mathrm{SVJ}}$. The sizes of the formula for robust test generation using $\eta_{\mathscr{L}_{19s}}$ is given in the last column for reason of comparison.

Note that not all gates have to be included in Ψ_{SVJ}. For example, if a rising transition occurs at the on-path input of an AND gate under the robust sensitization criterion, the side inputs – and consequently their fanin cone – do not have to be considered. This is because no static value has to be guaranteed here according to the sensitization criterion. Let G_S^F be the set of gates on which static values must be guaranteed for the PDF F. Then, only those gates that are located in the fanin cone of at least one gate $g \in G_S^F$ are included in Ψ_{SVJ}. As a result, the size of the formula for robust test generation can be reduced.

Table 8.8 CNF for an AND gate using $\eta_{\mathscr{L}_{4B}}$

$(\overline{x}_a^1 + \overline{x}_b^1 + x_c^1)$	$(x_a^1 + \overline{x}_c^1)$	$(x_b^1 + \overline{x}_c^1)$
$(\overline{x}_a^2 + \overline{x}_b^2 + x_c^2)$	$(x_a^2 + \overline{x}_c^2)$	$(x_b^2 + \overline{x}_c^2)$

Table 8.9 CNF description for static value justification for an AND gate using $\eta_{\mathscr{L}_{4B}}$

$(\overline{x}_a^2 + x_b^1 + \overline{x}_b^2 + \overline{x}_c^S)$	$(x_a^1 + \overline{x}_a^2 + \overline{x}_b^1 + \overline{x}_c^S)$	$(x_a^S + \overline{x}_b^2 + \overline{x}_c^S)$
$(\overline{x}_a^1 + \overline{x}_b^1 + x_b^2 + \overline{x}_c^S)$	$(x_a^S + \overline{x}_b^1 + \overline{x}_c^S)$	$(\overline{x}_a^2 + x_b^S + \overline{x}_c^S)$
$(\overline{x}_a^1 + x_a^2 + \overline{x}_b^2 + \overline{x}_c^S)$	$(\overline{x}_a^1 + x_b^S + \overline{x}_c^S)$	$(x_a^S + x_b^S + \overline{x}_c^S)$

Example 8.5. Consider an AND gate $c = a \cdot b$ with inputs a, b. The CNF that models an AND gate under $\mathscr{L}_{4B}$ is shown in Table 8.8. Given the additional variables x_a^S, x_b^S, x_c^S, the implications Ψ_{SVJ} described in (8.1) and (8.2) are presented in the CNF in Table 8.9. Each clause represents an illegal assignment of the variables. For example, the clause $(\overline{x}_a^1 + x_b^S + \overline{x}_c^S)$ implies that if a static value has to be justified at line c ($x_c^S = 1$) and the signal on line b is not guaranteed to be static ($x_b^S = 0$), line a has to assume the value 1 in the first time frame ($x_a^1 = 1$); see the Boolean encoding given in Table 8.7.

Note that each of these additional clauses contains the literal $\overline{x}_c^S$. This means that these clauses are only to be considered if $x_c^S = 1$, i.e. a static value has to be justified at this line. Otherwise, if $x_c^S = 0$, all clauses are satisfied by this assignment and no implication is derived.

8.5.3 Considering the Presence of Small Delay Defects

The distribution of *Small Delay Defects* (SDDs) has significantly increased in the last years due to the shrinking feature sizes. The growing distribution of SDDs results in an increased likeliness of delay defects on long paths [GH04, KMGE04] due to their possible accumulation as described in Sect. 7.2. The ARAP test generation procedure described above considers all side inputs in the same manner. No side input is prioritized. However, this is not optimal if SDDs are considered. This section shows how the presence of SDDs can be incorporated into the proposed test generation procedure. Note that the aim of the proposed method is not to detect SDDs but to avoid that tests are invalidated by SDDs.

Crucial in this incorporation is the consideration of path lengths. When the maximum number of static side inputs is the only objective of the generation of ARAP tests, some longer paths ending at side inputs can be ignored in favor of short paths ending at side inputs. Since shorter paths are typically easier to justify than longer paths, the test generation procedure is more likely to set static values on side inputs with short paths than on side inputs with long paths. However, the likeliness that a test is invalidated by a delay defect caused by SDDs is higher if a longer path ends at a side input. Therefore, side inputs with longer paths should be prioritized. Consequently, it is desired that static values should be set to side inputs with longer paths.

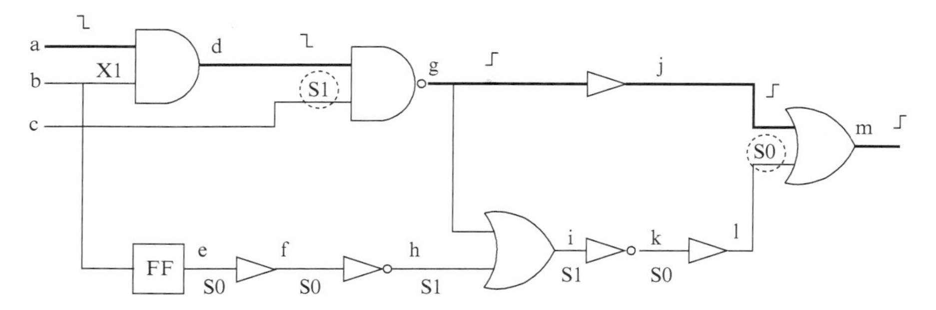

Fig. 8.5 ARAP test for path $a - d - g - j - m$ considering SDDs

Example 8.6. Figure 8.5 gives the example circuit which is also shown in Fig. 8.4 and used in Example 8.3. A test is shown which also sets a static value to two of the three side inputs. However, this test avoids its possible invalidation at the side input l which is set to a static value. In the presence of SDDs, this test is to be preferred since the side input with longer paths is covered by a static value. The test of Example 8.3 sets the same number of static side inputs. However, primary inputs are chosen only. The likeliness that SDDs cause delay defects on inputs is very low due to the short path length. Unfortunately, test generation algorithms are more likely to generate tests with static values on side inputs with short paths since their justification is typically easier to achieve.

In order to prioritize side inputs with longer paths, the PBO formulation is modified. The PBO instance $\Psi_{\mathscr{C}}^{F}$ for fault F in $\mathscr{C}$ is built in the same way as described in Sect. 8.5.1. However, the objective function $\mathscr{F}$ is formulated differently, since $\mathscr{F}$ is responsible for the maximization criterion. In the procedure described above, the constant (-1) used for each variable in $\mathscr{F}$ was responsible for the equal treatment of each side input. This constant is substituted by a value indicating the likeliness of the invalidation by SDDs.

Here, the length of the longest path ending at side input $s \in S_S^{\mathscr{P}}$ was chosen as an indicator. A static value on a long path avoids that the test can be invalidated via this path. The length is denoted by c_x for signal x in the following. The objective function is therefore formulated as follows:

$$\mathscr{F}(x_0^S, \ldots, x_n^S) = \sum_{i=0}^{n} (-c_{x_i}) \cdot x_i^S$$

By this, longer paths are prioritized since static values at the end of longer paths cause a higher maximum value in $\mathscr{F}$. If the PBO solver has to decide between two side inputs which cannot be set to static values at the same time, the one with the higher constant is chosen. The altered formulation of $\mathscr{F}$ is shown in the following example.

Example 8.7. In Fig. 8.5, the length of the longest path (under the unit delay model) for side input b and c is 1. Both side inputs are primary inputs. The length of side input l is 6 (path $e-f-h-i-k-l$). Therefore, $\mathscr{F}$ is formulated as follows:

$$\mathscr{F}(b^S, c^S, l^S) = (-1) \cdot b^S + (-1) \cdot c^S + (-6) \cdot l^S$$

Note that this is a heuristic measurement. Incorporating information about the path length leads to the possibility that a test is found which does not have the maximum number of static side inputs possible. However, experiments confirmed that the decrease in the number of static side inputs is negligible.

An advantage of using a PBO formulation is the flexibility of the objective function. Due to the objective function, the side inputs are dynamically ordered. In this book, we used the path length as indicator for demonstrating the efficiency and feasibility of the technique. This value can easily be replaced by more technology-dependent parameters or *Standard Delay Format* (SDF) information, respectively. Additionally, it is easily possible to extend the formulation to generate high-quality non-robust tests [CC94]. Here, the side inputs are statically ordered according to the difference between the arrival time of the on-input of the path under test and the earliest arrival time of the side input.

8.6 Experimental Results

The experimental results for SAT-based ATPG for the PDFM are presented in this section. The algorithms were implemented in C++ as the tool MONSOON. The tool MONSOON was integrated into the ATPG framework of NXP Semiconductors as a prototype using the SAT solver MiniSat [ES04] as well as DynamicSAT. Additionally, the ARAP test generation procedure was implemented using the PBO solver clasp [GKNS07]. More information about the circuits used and the experimental setup can be found in Sect. 5.7.3. The 40,000 longest paths of each circuit were chosen for PDF test generation.[5] The fixed delay model was applied for path extraction. For sake of simplicity, each gate has a fixed delay of 1. A broadside test (launch-on-capture) where the first test vector can be fully scanned is generated for each targeted PDF.

First, Sect. 8.6.1 shows a run time comparison with a competitive approach on Boolean circuits. In Sect. 8.6.2, the benefits of the SAT instance generation flow using several multiple-valued logics presented in Sect. 8.4 are evaluated for robust test generation. Section 8.6.3 presents the results of MONSOON using DynamicSAT and the circuit-based dynamic learning technique (see Chaps. 5 and 6). Experimental results for the ARAP test generation approach are given in Sect. 8.6.4.

---

[5]Note that results for p49k and p77k are not reported since no or only few testable paths could be found. The results for these circuits are therefore not meaningful.

Table 8.10 Run time comparison with competitive approach – non-robust

Circ.	KF-ATPG(min)	MONSOON(min)	Imp.
s5378	0:22	0:03	7.3x
s9234	0:10	0:04	2.5x
s13207	0:15	0:10	1.5x
s15850	0:20	0:08	2.5x
s35932	0:05	0:01	5.0x
s38417	0:33	0:06	5.5x
s38584	0:21	0:08	2.6x
b14	1:39	0:26	3.8x
b15	2:11	0:30	4.4x
b17	2:18	0:27	5.1x
b18	3:29	1:50	1.9x
b20	2:34	1:23	1.9x
b21	2:42	1:30	1.8x
b22	3:47	1:57	1.9x
Average			3.4x

8.6.1 Comparison with Competitive Approach

The efficiency of the CNF-based approach MONSOON for non-robust PDF test generation is shown in this section. Table 8.10 shows a run time comparison with KF-ATPG [YCW04] – a state-of-the-art circuit-SAT based PDF ATPG tool – which uses structural ATPG techniques. Furthermore, KF-ATPG maintains a path-status graph and, by this, works in an incremental manner. Since KF-ATPG is only able to generate non-robust tests for Boolean circuits, the run time comparison is performed only for this type of tests although the benefit of MONSOON lies in robust test generation for industrial circuits.[6]

Larger circuits from the ISCAS'89 as well as from the ITC'99 benchmark suite are taken as benchmarks. Column *Circ.* shows the circuit's name. Run time is given in CPU minutes. Results for KF-ATPG are presented in column *KF-ATPG*, whereas results for the proposed approach are given in column *MONSOON*. The factor of improvement of MONSOON is shown in column *Imp.* MONSOON clearly outperforms KF-ATPG by a factor of up to 7.3. The average factor of improvement is 3.4.

8.6.2 SAT Instance Generation Flow

Table 8.11 repeats some statistical information about the industrial circuits (see Table 5.7) and the last four columns show the results of the structural classification

[6]To the best of our knowledge, no PDF test generator that can generate robust tests and/or can handle static values and industrial constraints is available as public domain for comparison.

Table 8.11 Information about the industrial circuits

Circ.	#PIs	#POs	#FFs	LC_Z	LC_{U1}	LC_{U2}	LC_B
p44k	739	56	2,175	0	0	0	100
p57k	8	19	2,291	<0.1	0.2	25.7	74.0
p80k	152	75	3,878	0	0	0	100
p88k	403	183	4,302	0.6	7.3	21.4	70.7
p99k	167	82	5,747	0	1.5	5.6	92.9
p177k	768	1	10,507	0.8	27.4	54.5	17.3
p456k	1,723	72	14,900	3.3	20.4	75.7	0.6
p462k	1,815	604	29,205	0.2	13.6	34.7	51.5
p565k	996	201	32,409	6.5	7.3	59.0	27.2
p1330k	617	90	104,630	<0.1	9.0	15.0	75.9
p2787k	46,015	274	58,835	0.3	39.4	25.5	34.8
p3327k	4,093	274	148,184	3.5	16.1	80.2	0.2
p3852k	6,052	303	173,738	1.5	15.6	82.5	0.4

presented in Sect. 8.4. The percentage of gates contained in the identified logic class is given in each of these columns. Note that the number of gates in logic class LC_{U2} is larger than the number of gates in LC_{U1}, because unknown values reaching a flip-flop after the first time frame are propagated again in the second time frame. The large circuits especially contain a large number of gates which are not in the Boolean logic class LC_B. For example, only 0.2% of all gates in p3327k can be modeled in Boolean logic. On the other hand, the percentage of gates in LC_Z is also very small.

In order to show the benefit of the SAT instance generation flow presented in Sect. 8.4, a run time comparison of robust generation with and without the flow is given in the following. Table 8.12 presents the run time results for robust PDF test generation. The approach without incorporating the analysis (column *Classic*) handles all gates in the highest-valued logic $\mathcal{L}_{11s}/\mathcal{L}_{19s}$. Columns named *Av.#Cls* give the average number of clauses contained in the SAT instance. The number of aborts, i.e. PDFs which could not be classified in the given interval, is shown in column *Ab.* and the run time in CPU minutes (*min*) or CPU hours (*h*) is listed in columns named *Time*. The overall improvement in terms of run time is given in column *Imp.*

The results clearly show that the novel flow (column *Novel*) is mandatory for efficient SAT-based PDF test generation. The highest speed-up factor is greater than 35 (p44k). Parts of the improvement stems from run time savings during CNF generation. The CNF is larger without structural analysis which influences the run time needed to build the SAT instances. Without structural analysis, all gates are conservatively modeled as belonging to LC_Z/LC_{U1} – even gates belonging to LC_B. The novel flow unveils this overhead and gates are modeled more compactly. This is directly reflected by the overall size of the CNFs. With structural analysis, gates in LC_B are modeled by less clauses than gates in LC_Z/LC_{U1}. The SAT instances are therefore generally easier to solve. The number of unclassified faults also decreases significantly. Only 12% of the previously unclassified PDFs are still unclassified using the presented flow.

Table 8.12 Robust path delay test generation

	Classic			Novel			
Circ.	Av.#Cls	Ab.	Time	Av.#Cls.	Ab.	Time	Imp.
p44k	667,106	12,799	>72 h	304,889	0	2:02 h	>35.41x
p57k	308,927	1,259	3:52 h	130,065	898	1:40 h	2.32x
p80k	198,974	13,844	20:33 h	73,356	901	1:01 h	20.21x
p88k	98,930	0	40:08 min	41,134	0	8:30 min	4.72x
p99k	55,966	5	22:59 min	24,078	1	6:12 min	3.71x
p177k	516,722	9,338	39:46 h	307,409	3,042	10:17 h	3.87x
p456k	191,887	1,107	3:34 h	118,519	164	1:32 h	2.33x
p462k	93,331	0	1:00 h	55,857	0	39:49 min	1.51x
p565k	46,219	293	47:40 min	25,602	78	27:53 min	1.71x
p1330k	128,105	0	1:05 h	120,745	0	1:09 h	0.94x
p2787k	379,771	613	5:07 h	291,814	55	3:13 h	1.59x
p3327k	213,275	665	8:11 h	105,472	23	1:34 h	5.22x
p3852k	384,434	2,855	19:14 h	189,187	72	4:14 h	4.54x
Total		42,778			5,234		

8.6.3 MONSOON Using DynamicSAT

The experiments of MONSOON presented above are conducted with the SAT solver MiniSat as the underlying solving engine. MiniSat was replaced by DynamicSAT+ (Hybrid) for the experimental evaluation presented in this section. Here, strategy *Size 10* is used since it provided the best results in a preliminary evaluation.

Table 8.13 shows the experimental results for non-robust test generation and Table 8.14 presents the results for robust test generation. Column *MONSOON* gives the results using the underlying solving engine MiniSat and column *MONSOON+* presents the results for the approaches using DynamicSAT+, DynamicSAT+ Hybrid and MiniSat with circuit-based dynamic learning.[7]

Similar to the observation made for the SAFM and the TFM, the application of dynamic learning results in a significant decreased number of unclassified faults. Likewise to the other fault models, the technique is most effective for MONSOON+ using MiniSat. Here, the total number of aborts goes down from 10,382 to 150 which corresponds to only 1.4% of the previously unclassified PDFs. In particular important is that the run time decreases in most cases, too. For p177k, the factor of improvement is even 12.7. In contrast, the run time of MONSOON+ using DynamicSAT+ increases significantly for the same circuit due to the large amount of learned information. Although DynamicSAT+ is faster for most circuits, it has some outliers, i.e. p177k and p456k, where the run time is significantly higher than with MiniSat. Therefore, MONSOON+ using MiniSat is the most robust approach while MONSOON+ using DynamicSAT+ is the fastest approach.

[7]The version of MiniSat with circuit-based dynamic learning integrated is not named MiniSat+ but only MiniSat, since MiniSat+ is a pseudo-Boolean SAT solver [ES06].

Table 8.13 Experimental results – MONSOON using DynamicSAT – non-robust

	MONSOON		MONSOON+					
	MiniSat		MiniSat		DynamicSAT+		DynSAT+ Hyb.	
Circ.	Ab.	Time	Ab.	Time	Ab.	Time	Ab.	Time
p44k	0	1:35 h	0	1:44 h	0	1:06 h	0	1:16 h
p57k	7	46:00 min	5	20:35 min	27	15:29 min	12	14:39 min
p80k	16	49:25 min	2	1:24 h	49	2:41 h	28	2:08 h
p88k	0	6:39 min	0	5:41 min	0	2:58 min	0	2:59 min
p99k	0	5:30 min	0	5:07 min	1	6:23 min	0	6:25 min
p177k	8,274	39:22 h	9	3:06 h	336	14:39 h	74	15:38 h
p456k	652	2:09 h	110	1:37 h	244	7:00 h	207	9:14 h
p462k	0	33:06 min	0	32:09 min	0	18:08 min	0	18:21 min
p565k	1	25:03 min	0	37:11 min	53	48:05 min	8	37:14 min
p1330k	0	58:14 min	0	55:24 min	0	20:40 min	0	20:44 min
p2787k	473	4:08 h	23	2:59 h	31	1:23 h	112	3:44 h
p3327k	94	4:31 h	1	3:00 h	6	2:00 h	11	3:00 h
p3852k	865	9:10 h	0	3:41 h	2	1:52 h	4	1:57 h
Total	10,382		150		749		456	

Table 8.14 Experimental results – MONSOON using DynamicSAT – robust

	MONSOON		MONSOON+					
	MiniSat		MiniSat		DynamicSAT+		DynSAT+ Hyb.	
Circ.	Ab.	Time	Ab.	Time	Ab.	Time	Ab.	Time
p44k	0	2:02 h	0	1:03 h	0	2:08 h	0	3:11 h
p57k	898	1:40 h	20	48:38 min	234	4:21 h	99	3:25 h
p80k	901	1:01 h	80	1:10 h	292	1:39 h	227	1:25 h
p88k	0	8:30 min	0	7:09 min	0	3:45 min	0	3:45 min
p99k	1	6:12 min	0	6:33 min	0	4:45 min	0	4:49 min
p177k	3,042	10:17 h	70	3:04 h	672	24:15 h	264	43:01 h
p456k	164	1:32 h	9	1:24 h	83	1:33 h	47	1:32 h
p462k	0	39:49 min	0	31:43 min	2	18:35 min	0	15:22 min
p565k	78	27:53 min	34	40:55 min	102	41:49 min	67	39:24 min
p1330k	0	1:09 h	0	58:58 min	0	21:01 min	0	21:27 min
p2787k	55	3:13 h	10	2:40 h	4	55:22 min	8	58:07 min
p3327k	23	1:34 h	1	2:21 h	12	1:14 h	6	1:22 h
p3852k	72	4:14 h	2	3:06 h	18	1:47 h	8	2:09 h
Total	5,234		226		1,419		726	

The experiments for robust test generation show a slightly different picture. As already observed for non-robust test generation, the circuit-based dynamic learning technique is able to significantly reduce the number of aborts for robust test generation which is more complex. Here, MONSOON+ using MiniSat is the most robust approach producing very few aborts. The reduction rate, i.e. the percentage of the classified faults, which were aborted before, of MONSOON+ using MiniSat is 4.3%. However, MONSOON+ using DynamicSAT+ is faster but – likewise to

Table 8.15 Experimental results – run time

Circ.	#Paths	#NR	%Rob.	Run time CNF	ARAP	SDD
b14	1,666	354	1.7	0:36 min	1:00 min	1:46 min
b15	3,696	346	0	1:23 min	1:32 min	1:35 min
b17	11,396	2,427	3.0	6:06 min	8:23 min	11:33 min
b18	29,214	12,198	2.5	33:29 min	59:37 min	1:23 h
b20	4,646	1,648	7.7	4:36 min	9:39 min	15:27 min
b21	4,510	1,485	8.5	4:07 min	9:33 min	16:52 min
b22	6,716	2,235	9.5	5:44 min	12:17 min	19:19 min
p57k	16,042	4,142	45.9	1:17 h	2:39 h	2:57 h
p80k	22,062	15,034	5.4	1:49 h	3:48 h	4:26 h
p99k	12,710	7,095	6.8	8:25 min	23:14 min	31:25 min
p462k	40,000	8,855	19.0	6:01 min	13:59 min	37:07 min
p565k	40,000	9,909	14.2	15:51 min	28:35 min	36:07 min
p1330k	40,000	8,473	80.2	1:09 min	3:12 min	8:02 min
p3327k	40,000	17,851	22.0	45:07 min	1:17 h	2:09 h

non-robust test generation – has some outliers, i.e. p57k and p177k, where the run time is not acceptable compared to MONSOON+ using MiniSat. The reason for the outliers of DynamicSAT+ is the large number of learned clauses which slow down the search process significantly. Therefore, the first choice is MONSOON+ using MiniSat.

Overall, the experiments have shown that SAT-based ATPG is very efficient for PDF test generation. The proposed SAT-based ATPG techniques are able to cope with the complexity of robust test generation producing a very low number of aborts.

8.6.4 ARAP Test Generation

This section provides experimental results of the presented ARAP test generation approach. The experiments were conducted on ITC'99 benchmark circuits as well as on industrial circuits.

Table 8.15 shows the detailed results of the test generation which was performed using the launch-on-capture scheme. Column *Circ.* gives the name of the circuit. For each circuit, 40,000 long critical paths with a path length of over 50 elements were extracted (if exists). The total number of extracted paths is given in column *#Paths*. The number of non-robustly testable PDFs is presented in column *#NR*. Column *%Rob.* presents the percentage of robustly testable PDFs among the non-robustly testable PDFs.

Three different approaches were evaluated. The run time of these approaches is given in column *Run time* in CPU minutes (*min*) or CPU hours (*h*). Column *CNF* gives the run time of CNF-based test generation for generating robust as well as non-robust tests. However, no ARAP test is generated. Column *ARAP* gives the

Table 8.16 Experimental results – ARAP tests

	ARAP tests			Length	
Circ.	%ARAP	%St_Inp.	%Min_Inp.	Av.Imp.(%)	Max.Imp.(%)
b14	98.3	89.3	15.2	+0	+0
b15	100.0	83.5	8.6	+0	+0
b17	97.0	85.7	5.7	+2	+11
b18	97.5	91.8	50.0	+8	+130
b20	92.3	92.6	41.9	+1	+90
b21	91.5	91.4	41.9	+1	+108
b22	90.5	91.9	30.2	+1	+7
p57k	54.1	87.0	1.0	+2	+249
p80k	94.6	86.9	11.5	+12	+887
p99k	93.2	74.9	5.3	+5	+46
p462k	81.0	87.7	8.3	+6	+99
p565k	85.8	81.8	31.3	+4	+92
p1330k	19.8	83.5	39.3	+10	+70
p3327k	78.0	88.9	1.0	+4	+66

run time for the proposed ARAP test generation (including robust tests). The run time overhead for generating ARAP tests is in most cases only about twice as long and ranges between factors of 1.1 (b15) and 2.3 (b21) for the benchmark circuits and between factors of 1.7 (p3327k) and 2.8 (p1330k) for the industrial circuits. In contrast, the quality of the tests significantly increases.

Table 8.16 gives the results concerning the test quality. Column *%ARAP* reports the percentage of generated ARAP tests. In fact, each of the robustly untestable faults could be tested by an ARAP test since at least one input could be set to a static value. The average percentage of side inputs which could be set to a static value is given in column *%St_Inp*. The results show that the average percentage is very high ranging between 74–92%. Most of the side inputs can be set to a static value. This confirms the large quality gap between non-robust and robust tests. The minimum percentage of static side inputs is also given in column *%Min_Inp*.

The run time for ARAP test generation considering SDDs is presented in column *SDD*. Here, the run time overhead is higher compared to ARAP test generation. The overhead ranges between factors of 1.1 (b15) and 3.6 (b21) for the benchmark circuits and between factors of 2.3 (p57k) and 6.9 (p1330k) for the industrial circuits. This is due to the fact, that more solutions have to be considered because of the incorporation of different factors for each side input. In order to measure the quality improvement of this method, all side inputs set to a static value are considered in each test and the average path length of longest paths ending at these side inputs is calculated. On these paths, accumulated SDDs are not able to invalidate the test. The improvement of the average path length is presented in column *Av.Imp.* The results show that the average path length with static values is increased on average by up to 12%. Column *Max.Imp.* gives the highest improvement which could be achieved for one test. The highest factor of improvement can be obtained for p80k with an increase of 887% for one test.

In summary, the experiments confirm the large quality gap between non-robust and robust tests which can be significantly diminished by the ARAP test generation approach. The results show that the generation of ARAP tests causes moderate run time overhead compared to non-robust/robust test generation. However, it is also shown that ARAP increases the quality of the generated tests significantly since ARAP tests contain a very high average number of static inputs. The extended formulation considering the presence of SDDs is able to increase the average path length of side inputs with static values. This decreases the likeliness that the test is invalidated by delay defects caused by SDDs.

8.7 Summary

The *Path Delay Fault Model* (PDFM) is the most accurate delay fault model and tests for *Path Delay Faults* (PDFs) are widely used for guaranteeing the correctness of critical paths or for diagnostic reasoning. Therefore, high quality tests, i.e. robust tests, are highly important for PDF test generation.

This chapter has introduced MONSOON, a SAT-based approach for generating non-robust and robust tests for PDFs in an industrial environment. The modeling of static values is necessary for robust test generation. Furthermore, the industrial application requires the consideration of additional values. A 19-valued logic has been used in order to model these requirements and the transformation into a Boolean SAT problem using a Boolean encoding is described.

The use of a 19-valued logic already signifies a huge overhead for the size of the SAT problem and the ATPG run time. Therefore, a SAT instance generation flow has been presented which makes use of structural properties of the circuit. As a result, the circuit's elements are inserted into logic classes, which are mapped onto different multiple-valued logics with fewer values according to the desired quality of the test. The use of different multiple-valued logics has been shown to be mandatory for efficient SAT-based PDF test generation since it is able to reduce the size of the SAT instance significantly.

Besides the general procedure to efficiently generate robust tests, a new quality level for robustly untestable faults has been introduced. Robustly untestable paths are typically tested with non-robust tests. However, there is a large quality gap between robust and non-robust tests. The quality level of *As-Robust-As-Possible* tests has been presented which is able to diminish this quality gap. A test generation algorithm based on *Pseudo-Boolean Optimization* (PBO) has been introduced which is able to generate tests with highest quality possible in one step. The proposed approach is able to incorporate the presence of *Small Delay Defects* (SDDs) in order to reduce the likelihood that tests are invalidated by these defects. This shows the flexibility of the formulation as a PBO problem.

Experiments on industrial circuits have shown that MONSOON is able to cope with the problem of high-quality PDF test generation. The proposed SAT techniques allow for an efficient and robust test generation process for generating high-quality tests in an industrial environment with very few aborts. It has also been shown that the SAT solving techniques proposed in Chaps. 5 and 6 are able to improve PDF test generation as well.

Chapter 9
Summary and Outlook

The production test is an important task in the design and manufacturing flow of today's circuits. Test patterns generated by *Automatic Test Pattern Generation* (ATPG) algorithms ensure that erroneous chips are filtered out before being delivered to customers. However, classical ATPG algorithms reach their limit and the high fault coverage demands of the industry are compromised since these algorithms have difficulties to cope with the growing number of hard-to-detect faults. Due to the shrinking feature sizes and increased speed of modern designs, delay fault testing becomes more and more important. The problem of the large number of unclassified faults is very serious in delay fault test generation which is more complex than classical stuck-at fault test generation.

Algorithms based on the *Boolean Satisfiability* (SAT) problem have been shown to be robust in the field of ATPG. SAT-based ATPG was introduced in the early 1990s but could not be established in industrial practice due to several shortcomings. These shortcomings have been addressed in this book. Focus of the research work was the efficient application of SAT-based ATPG in an industrial environment. Algorithmic improvements have been presented in order to improve the performance and robustness of SAT-based ATPG for large industrial circuits.

The proposed SAT technique *Dynamic Clause Activation* (DCA) improves the performance of SAT-based ATPG by reducing the SAT transformation time and applying structural knowledge. The run time gap between classical ATPG algorithms and SAT-based ATPG can be significantly diminished or even closed using this technique. The robustness of SAT-based ATPG can be further increased by using the proposed circuit-based dynamic learning technique. This is especially important for the complex task of delay fault test generation. The combination of both techniques provides an efficient ATPG engine which is fast and, at the same time, very robust. The drawback of the proposed SAT techniques is the memory consumption for large circuits. This has to be addressed in future work by, e.g. circuit partitioning techniques or interval-based clause deletion.

The resulting ATPG engine has been extensively evaluated in an industrial test environment on industrial circuits containing multi-million elements. The

S. Eggersglüß and R. Drechsler, *High Quality Test Pattern Generation and Boolean Satisfiability*, DOI 10.1007/978-1-4419-9976-4_9,

evaluations have shown that SAT-based ATPG using the techniques presented in this book is well suitable for an efficient application in industrial practice and reduces the number of unclassified faults to a minimum. This results in a significant fault coverage increase of up to 2% for the transition fault model. The integration of the presented SAT-based ATPG engine clearly enhances the industrial test environment.

Additionally, special attention was paid to the generation of high-quality tests for delay faults since the demands for these types of tests grows. Delay tests can traditionally be classified into non-robust and robust tests. Robust tests are more desirable but also harder to obtain. A SAT instance generation flow for robust tests based on the use of logic classes and several multiple-valued logics has been shown. This flow uses a structural analysis of the circuit to reduce the SAT instance size significantly. As a result, the ATPG process can be significantly accelerated.

Many defects occurring in practice, i.e. small delay defects, are not covered by the classical fault models anymore. Therefore, the demands for high-quality tests or tests for newly developed fault models are increasing. Often, test generation for these types of defects involves additional information such as timing. These types of information cannot or only scarcely be integrated in a Boolean search process. One feasible solution is the application of solvers for *Pseudo-Boolean Optimization* (PBO) which are able to process additional information in a specific way. In this book, the application and the effectiveness of PBO solvers in the field of test generation has been demonstrated.

As-Robust-As-Possible tests and a corresponding PBO-based ATPG algorithm have been presented in order to close the quality gap between non-robust and robust tests. Furthermore, it has been shown how the timing-aware ATPG problem can be encoded as a PBO problem in order to efficiently generate high-quality transition tests for small delay defects.

The application of PBO solvers to increase the quality of the generated tests has been shown to be very promising. By this, more information about the problem can be incorporated into the search process, and at the same time, efficient SAT techniques can be used to find a solution. However, instead of terminating the search after finding one solution, PBO-based algorithms have to search the best among all solutions which means a large overhead for the search process. Therefore, more sophisticated solving algorithms are needed to cope with the increased complexity of optimization problems. A future direction of research work is to develop PBO solving algorithms which are able to use knowledge about the ATPG problem to explore the solution space much faster. A possible approach could be to learn non-Boolean dependencies at the circuit level, e.g. with respect to timing information, and reuse them in subsequent problem instances.

Another possible field of research is to the leverage the possibilities of multi-cores. ATPG algorithms are mainly single-threaded and, consequently, do not use the full computational power of multi-cores. A possible application is the use of parallelization in pre-processing the circuit, i.e. learning logic dependencies. Learning implications in a pre-processing step takes much time for large circuits. By parallelizing this step, more powerful implications can be learned which have the potential to accelerate the search process.

References

ABF90. Abramovici M, Breuer MA, Friedman, AD (1990) Digital systems testing and testable design. Computer Science Press, New York, USA

ATJ06. Ahmed N, Tehranipoor M, Jayram V (2006) Timing-based delay test for screening small delay defects. In: Proceedings of the design automation conference, pp 320–325

Anj09. Anjos M (2009) Pseudo-Boolean forms. In: Biere A, Heule M, Maaren Hv, Walsh T (eds) Handbook of satisfiability. Frontiers in artificial intelligence and applications. IOS Press, Amsterdam, The Netherlands, pp 49–51

BR83. Barzilai Z, Rosen BK (1983) Comparison of AC self-testing procedures. In: Proceedings of the international test conference, pp 89–94

Bec98. Becker B (1998) Testing with decision diagrams. Integr VLSI J 26(1–2):5–20

BAA92. Bhattacharya D, Agrawal P, Agrawal VD (1992) Delay fault test generation for scan/hold circuits using Boolean expressions. In: Proceedings of the design automation conference, pp 159–164

Bie08a. Biere A (2008a) Adaptive restart strategies for conflict driven SAT solvers. In: Proceedings of the international conference on theory and applications of satisfiability testing. Volume 4996 of lecture notes in computer science, pp 28–33

Bie08b. Biere A (2008b) PicoSAT essentials. J Satisfiability, Boolean Model Comput 4(2–4):75–97

BAA98. Bose S, Agrawal P, Agrawal VD (1998) Deriving logic systems for path delay test generation. IEEE Trans Comput 47(8):829–846

Bra93. Brand D (1993) Verification of large synthesized designs. In: Proceedings of the international conference on computer-aided design, pp 534–537

BF85. Brglez F, Fujiwara H (1985) A neutral netlist of 10 combinational circuits and a target translator in fortran. In: Proceedings of the IEEE international symposium on circuits and systems, pp 663–698

BBK89. Brglez F, Bryan D, Kozminski K (1989) Combinational profiles of sequential benchmark circuits. In: Proceedings of the IEEE international symposium on circuits and systems, pp 1929–1934

Bry86. Bryant RE (1986) Graph-based algorithms for Boolean function manipulation. IEEE Trans Comput 35(8):677–691

BA00. Bushnell ML, Agrawal VD (2000) Essentials of electronic testing for digital, memory and mixed-signal VLSI circuits. Kluwer, Boston, USA

CAR93. Chakradhar ST, Agrawal VD, Rothweiler SG (1993) A transitive closure algorithm for test generation. IEEE Trans Comput-Aided Des Integr Circuits Syst 12(7):1015–1028

S. Eggersglüß and R. Drechsler, *High Quality Test Pattern Generation and Boolean Satisfiability*, DOI 10.1007/978-1-4419-9976-4,

CH05. Chandrasekar K, Hsiao MS (2005) Integration of learning techniques into incremental satisfiability for efficient path-delay fault test generation. In: Proceedings of design, automation and test in Europe, pp 1002–1007

CG96. Chen C, Gupta SK (1996) A satisfiability-based test generator for path delay faults in combinational circuits. In: Proceedings of the design automation conference, pp 209–214

Che93. Cheng K-T (1993) Transition fault testing for sequential circuits. IEEE Trans Comput-Aided Des Integr Circuits Syst 12(12):1971–1983

CC94. Cheng K-T, Chen H-C (1994) Generation of high-quality non-robust tests for path delay faults. In: Proceedings of the design automation conference, pp 365–369

Coo71. Cook SA (1971) The complexity of theorem proving procedures. In: ACM symposium on theory of computing, pp 151–158

CSS00. Corno F, Sonza-Reorda M, Squillero G (2000) RT-level ITC 99 benchmarks and first ATPG results. In: IEEE design & test of computers, pp 44–53

CPE$^+$09. Czutro A, Polian I, Engelke P, Reddy SM, Becker B (2009) Dynamic compaction in SAT-based ATPG. In: Proceedings of the IEEE Asian test symposium, pp 187–190

CPL$^+$10. Czutro A, Polian I, Lewis M, Engelke P, Reddy SM, Becker B (2010) Thread-parallel integrated test pattern generator utilizing satisfiability analysis. Int J Parallel Progr 38(3–4):185–202

DP60. Davis M, Putnam H (1960) A computing procedure for quantification theory. J ACM 7(3):201–215

DLL62. Davis M, Logemann G, Loveland D (1962) A machine program for theorem proving. Commun ACM 5(7):394–397

DS91. Dervisoglu BI, Stong GE (1991) Design for testability: using scanpath techniques for path-delay test and measurement. In: Proceedings of the international test conference, pp 365–374

Dre94. Drechsler R (1994) BiTeS: a BDD based test pattern generator for strong robust path delay faults. In: Proceedings of the European conference on design automation, pp 322–327

DEF$^+$08. Drechsler R, Eggersglüß S, Fey G, Glowatz A, Hapke F, Schloeffel J, Tille D (2008) On acceleration of SAT-based ATPG for industrial designs. IEEE Trans Comput-Aided Des Integr Circuits Syst 27(7):1329–1333.

DEFT09. Drechsler R, Eggersglüß S, Fey G, Tille D (2009) Test pattern generation using boolean proof engines. Springer, Heidelberg, Germany

EB05. Eén N, Biere A (2005) Effective preprocessing in SAT through variable and clause elimination. In: Proceedings of the international conference on theory and applications of satisfiability testing. Volume 3569 of lecture notes in computer science, pp 61–75

ES03. Eén N, Sörensson N (2003) Temporal induction by incremental SAT solving. In: Proceedings of the international workshop on bounded model checking. Volume 89 of electronic notes in theoretical computer science

ES04. Eén N, Sörensson N (2004) An extensible SAT solver. In: Proceedings of the international conference on theory and applications of satisfiability testing. Volume 2919 of lecture notes in computer science, pp 502–518

ES06. Eén N, Sörensson N (2006) Translating pseudo-Boolean constraints into SAT. J Satisfiability, Boolean Model Comput 2(1–4):1–26

ED07. Eggersglüß S, Drechsler R (2007) Improving test pattern compactness in SAT-based ATPG. In: Proceedings of the IEEE Asian test symposium, pp 445–450

ED08. Eggersglüß S, Drechsler R (2008) On the influence of Boolean encodings in SAT-based ATPG for path delay faults. In: Proceedings of the international symposium on multiple-valued logic, pp 94–99

ED09. EggersglüßS, Drechsler R (2009) Increasing robustness of SAT-based delay test generation using efficient dynamic learning techniques. In: Proceedings of the IEEE European test symposium, pp 81–86

ED11a. Eggersglüß S, Drechsler R (2011a) As-Robust-As-Possible test generation in the presence of small delay defects using pseudo-Boolean optimization. In: Proceedings of design, automation and test in Europe, pp 1291–1297

ED11b. Eggersglüß S, Drechsler R (2011b) Efficient data structures and methodologies for SAT-based ATPG providing high fault coverage in industrial application. IEEE Trans Comput-Aided Des Integr Circuits Syst 30(9):1411–1415

ED11c. Eggersglüß S, Drechsler R (2011c) On timing-aware ATPG using pseudo-Boolean optimization. In: IEEE European test symposium, informal digest of papers

ED12. Eggersglüß S, Drechsler R (2012) A highly fault-efficient SAT-based ATPG flow. IEEE Des Test Comput

ETF+07. Eggersglüß S, Tille D, Fey G, Drechsler R, Glowatz A, Hapke F, Schloeffel J (2007) Experimental studies on SAT-based ATPG for gate delay faults. In: Proceedings of the international symposium on multiple-valued logic

EFG+10. Eggersglüß S, Fey G, Glowatz A, Hapke F, Schloeffel J, Drechsler R (2010) MONSOON: SAT-based ATPG for path delay faults using multiple-valued logics. J Electron Test: Theory Appl 26(3):307–322

EW77. Eichelberger EB, Williams TW (1977) A logic design structure for design for testability. In: Proceedings of the design automation conference, pp 462–468

Eld59. Eldred RD (1959) Test routines based on symbolic logical statements. J ACM 6(1): 33–36

FSD06. Fey G, Shi J, Drechsler R (2006) Efficiency of multi-valued encoding in SAT-based ATPG. In: Proceedings of the international symposium on multiple-valued logic, pp 25–30

FWD07. Fey G, Warode T, Drechsler R (2007) Reusing learned information in SAT-based ATPG. In: Proceedings of the international conference on VLSI design, pp 69–76

FYM05. Fu Z, Yu Y, Malik S (2005) Considering circuit observability don't cares in CNF satisfiability. In: Proceedings of design, automation and test in Europe, pp 1108–1113

FFS91. Fuchs K, Fink F, Schulz MH (1991) DYNAMITE: an efficient automatic test pattern generation system for path delay faults. IEEE Trans Comput-Aided Des Integr Circuits Syst 10(10):1323–1335

FWA93. Fuchs K, Wittmann HC, Antreich KJ (1993) Fast test pattern generation for all path delay faults considering various test classes. In: Proceedings of the European test conference, pp 89–98.

FPR94. Fuchs K, Pabst M, Rössel T (1994) RESIST: a recursive test pattern generation algorithm for path delay faults considering various test classes. IEEE Trans Comput-Aided Des Integr Circuits Syst 13(12):1550–1562

FS83. Fujiwara H, Shimono T (1983) On the acceleration of test generation algorithms. IEEE Trans Comput 32(12):1137–1144

FT82. Fujiwara H, Toida S (1982) The complexity of fault detection problems for combinational logic circuits. IEEE Trans Compu 31(6):555–560

GNR61. Galey JM, Norby RE, Roth, JP (1961) Techniques for the diagnosis of switching circuit failures. In: Proceedings of the annual symposium on switching circuit theory and logical design, pp 152–160

GZA+02. Ganai MK, Zhang L, Ashar P, Gupta A, Malik S (2002) Combining strengths of circuit-based and CNF-based algorithms for a high-performance SAT solver. In: Proceedings of the design automation conference, pp 747–750

GKNS07. Gebser M, Kaufmann B, Neumann A, Schaub T (2007) Conflict-driven answer set solving. In: Proceedings of the international joint conference on artificial intelligence, pp 386–392

GBA97. Gharaybeh MA, Bushnell ML, Agrawal VD (1997) Classification and test generation for path-delay faults using single stuck-at fault tests. J Electron Test: Theory Appl 11(1):55–67

GB90. Giraldi J, Bushnell ML (1990) EST: the new frontier in automatic test-pattern generation. In: Proceedings of the design automation conference, pp 667–672

GF02. Gizdarski E, Fujiwara H (2002) SPIRIT: a highly robust combinational test generation algorithm. IEEE Trans Comp-Aided Des Integr Circuits Syst 21(12):1446–1458

Goe81. Goel P (1981) An implicit enumeration algorithm to generate tests for combinational logic. IEEE Trans Comput 30(3):215–222

GN02. Goldberg E, Novikov Y (2002) BerkMin: a fast and robust SAT-solver. In: Proceedings of design, automation and test in Europe, pp 142–149

GT80. Goldstein LH, Thigpen EL (1980) SCOAP: sandia controllability/observability analysis program. In: Proceedings of the design automation conference, pp 190–196

GSK98. Gomes CP, Selman B, Kautz H (1998) Boosting combinatorial search through randomization. In: Proceedings of the national conference on artificial intelligence, pp 431–437

GSM99. Guerra e Silva L, Silveira LM, Marques-Silva JP (1999) Algorithms for solving Boolean satisfiability in combinational circuits. In: Proceedings of design, automation and test in Europe, pp 526–530

GH04. Gupta P, Hsiao MS (2004) ALAPTF: a new transition fault model and the ATPG algorithm. In: Proceedings of the international test conference, pp 1053–1060

GGYA01. Gupta A, Gupta A, Yang Z, Ashar P (2001) Dynamic detection and removal of inactive clauses in SAT with application in image computation. In: Proceedings of the design automation conference, pp 536–541

HP99. Hamzaoglu I, Patel JH (1999) New techniques for deterministic test pattern generation. J Electron Test: Theory Appl 15(1–2):63–73

Hoo93. Hooker JN (1993) Solving the incremental satisfiability problem. J Logic Progr 15(1–2):177–186

HRVD77. Hsieh EP, Rasmussen RA, Vidunas LJ, Davis WT (1977) Delay test generation. In: Proceedings of the design automation conference, pp 486–491

IRS88. Iyengar VS, Rosen BK, Spillinger IY (1988) Delay test generation 2: algebra and algorithms. In: Proceedings of the international test conference, pp 867–876

IPC03. Iyer MK, Parthasarathy G, Cheng K-T (2003) SATORI – a fast sequential SAT engine for circuits. In: Proceedings of the international conference on computer-aided design, pp 320–325

JG03. Jha NK, Gupta S (2003) Testing of digital systems. Cambridge University Press, Cambridge

KMT+06. Kajihara S, Morishima S, Takuma A, Wen X, Maeda T, Hamada S, Sato Y (2006) A framework of high-quality transition fault ATPG for scan circuits. In: Proceedings of the international test conference

KWMS00. Kim J, Whittemore J, Marques-Silva JP, Sakallah KA (2000) On applying incremental satisfiability to delay fault testing. In: Proceedings of design, automation and test in Europe, pp 380–384

KM87. Kirkland T, Mercer MR (1987) A topological search algorithm for ATPG. In: Proceedings of the design automation conference, pp 502–508

KL93. Konuk H, Larabee T (1993) Explorations of sequential ATPG using Boolean satisfiability. In: Proceedings of the VLSI test symposium, pp 85–90

KC98. Krstić A, Cheng K-T (1998) Delay fault testing for VLSI circuits. Kluwer, Boston

KMGE04. Kruseman B, Majhi AK, Gronthoud G, Eichenberger S (2004) On hazard-free patterns for fine-delay fault testing. In: Proceedings of the international test conference, pp 213–222

KP93. Kunz W, Pradhan DK (1993) Accelerated dynamic learning for test pattern generation in combinational circuits. IEEE Trans Comput-Aided Des Integr Circuits Syst 12(5):684–694

KP94. Kunz W, Pradhan DK (1994) Recursive learning: a new implication technique for efficient solutions to CAD problems-test, verification, and optimization. IEEE Trans Comput-Aided Des Integr Circuits Syst 13(9):1143–1158

Lar92. Larrabee T (1992) Test pattern generation using Boolean satisfiability. IEEE Trans Comput-Aided Des Integr Circuits Syst 11(1):4–15

LH93. Lee HK, Ha DS (1993) Atalanta: an efficient ATPG for combinational circuit. Technical report, Department of Electrical Engineering, Virginia Polytechnic Institute and State University

LM86. Levendel Y, Menon PR (1986) Transition faults in combinational circuits: input transition test generation and fault simulation. In: Proceedings of the international symposium on fault-tolerant computing, pp 278–283

LSB07. Lewis M, Schubert T, Becker B (2007) Multithreaded SAT solving. In: Proceedings of the ASP design automation conference, pp 926–931

LR87. Lin C-J, Reddy SM (1987) On delay fault testing in logic circuits. IEEE Trans Comput-Aided Des Integr Circuits Syst 6(5):694–703

LRS89. Li W-N, Reddy SM, Sahni SK (1989) On path selection in combinational logic circuits. IEEE Trans Comput-Aided Des Integr Circuits Syst 8(1):56–63

LTW$^+$06. Lin X, Tsai K-H, Wang C, Kassab M, Rajski J, Kobayashi T, Klingenberg R, Sato Y, Hamada S, Aikyo T (2006) Timing-aware ATPG for high quality at-speed testing of small delay defects. In: Proceedings of the IEEE Asian test symposium, pp 139–146

vdL96. Linden Hvd (1996) Automatic test pattern generation for three-state circuits. Thesis Technische, University of Delft

Lio92. Lioy A (1992) Advanced fault collapsing. IEEE Des Test Comput 9(1):64–71

LWCH03. Lu F, Wang L-C, Cheng K-T, Huang R (2003a) A circuit SAT solver with signal correlation guided learning. In: Proceedings of design, automation and test in Europe, pp 892–897

LWC$^+$03. Lu F, Wang L-C, Cheng K-T, Moondanos J, Hanna Z (2003b) A signal correlation guided ATPG solver and its applications for solving difficult industrial cases. In: Proceedings of the design automation conference, pp 436–441

LHL07. Lu S-Y, Hsieh M-T, Liou J-J (2007) An efficient SAT-based path delay fault ATPG with an unified sensitization model. In: Proceedings of the international test conference, pp 1–7

Mah93. Mahlstedt U (1993) DELTEST: deterministic test generation for gate delay faults. In: Proceedings of the international test conference, pp 972–980

MGÖD90. Mahlstedt U, Grüning T, Özcan C, Daehn W (1990) CONTEST: a fast ATPG tool for very large combinational circuits. In: Proceedings of the international conference on computer-aided design, pp 222–225

MV07. Manolios P, Vroon D (2007) Efficient circuit to CNF conversion. In: Proceedings of the international conference on theory and applications of satisfiability testing. Volume 4501 of lecture notes in computer science, pp 4–9

MC92. Mao W, Ciletti MD (1992) Robustness enhancement and detection threshold reduction in ATPG for gate delay faults. In: Proceedings of the international test conference, pp 588–597

Mar99. Marques-Silva JP (1999) The impact of branching heuristics in propositional satisfiability algorithms. In: Proceedings of the Portuguese conference on artificial intelligence, pp 62–74

Mar00. Marques-Silva JP (2000) Algebraic simplification techniques for propositional satisfiability. In: Proceedings of the international conference on principles and practice of constraint programming. Volume 1894 of lecture notes in computer science, pp 537–542

MS94. Marques-Silva JP, Sakallah KA (1994) Dynamic search-space pruning techniques in path sensitization. In: Proceedings of the design automation conference, pp 705–711

MS97. Marques-Silva JP, Sakallah KA (1997) Robust search algorithms for test pattern generation. In: Proceedings of the international symposium on fault-tolerant computing, pp 152–157

MS99. Marques-Silva JP, Sakallah KA (1999) GRASP: a search algorithm for propositional satisfiability. IEEE Trans Comput 48(5):506–521

Moo65. Moore GE (1965) Cramming more components onto integrated circuits. Electronics 38(8):114–117

MMZ+01. Moskewicz MW, Madigan CF, Zhao Y, Zhang L, Malik S (2001) Chaff: engineering an efficient SAT solver. In: Proceedings of the design automation conference, pp 530–535

Mut76. Muth P (1976) A nine-valued circuit model for test generation. IEEE Trans Comput 25(6):630–636

PM92. Park ES, Mercer MR (1992) An efficient delay test generation system for combinational logic circuits. IEEE Trans Comput-Aided Des Integr Circuits Syst 11(7): 926–938

PRU95. Pomeranz I, Reddy SM, Uppaluri P (1995) NEST: a nonenumerative test generation method for path delay faults in combinational circuits. IEEE Trans Comput-Aided Des Integr Circuits Syst 14(12):1505–1515

PR88. Pramanick AK, Reddy SM (1988) On the detection of delay faults. In: Proceedings of the international test conference, pp 845–856

PG06. Putman R, Gawde R (2006) Enhanced timing-based transition delay testing for small delay defects. In: Proceedings of the VLSI test symposium, pp 336–342

RC90. Rajski J, Cox H (1990) A method to calculate necessary assignments in algorithmic test pattern generation. In: Proceedings of the international test conference, pp 25–34

Rot66. Roth JP (1966) Diagnosis of automata failures: a calculus and a method. IBM J Res Dev 10:278–281

RBS67. Roth JP, Bouricius WG, Schneider PR (1967) Programmed algorithms to compute tests to detect and distinguish between failures in logic circuits. IEEE Trans Electron Comput 16(5):567–580

SVDL04. Safarpour S, Veneris A, Drechsler R, Lee J (2004) Managing don't cares in Boolean satisfiability. In: Proceedings of design, automation and test in Europe, pp 260–265

SVD08. Safarpour S, Veneris A, Drechsler R (2008) Improved SAT-based reachability analysis with observability don't cares. J Satisfiability, Boolean Model Comput 5:1–25

SBS92. Saldanha A, Brayton RK, Sangiovanni-Vincentelli AL (1992) Equivalence of robust delay-fault and single stuck-fault test generation. In: Proceedings of the design automation conference, pp 173–176

SCS+11. Sauer M, Czutro A, Schubert T, Hillebrecht S, Polian I, Becker B (2011) SAT-based analysis of sensitisable paths. In: Proceedings of the IEEE symposium on design and diagnosis of electronic circuits and systems, pp 93–98

SP93. Savir J, Patil S (1993) Scan-based transition test. IEEE Trans Comput-Aided Des Integr Circuits Syst 12(8):1232–1241

SP94. Savir J, Patil S (1994) Broad-side delay test. IEEE Trans Comput-Aided Des Integr Circuits Syst 13(8):1057–1064

SA89. Schulz MH, Auth E (1989) Improved deterministic test pattern generation with applications to redundancy identification. IEEE Trans Comput-Aided Des Integr Circuits Syst 8(7):811–816

SB87. Schulz MH, Brglez F (1987) Accelerated transition fault simulation. In: Proceedings of the design automation conference, pp 237–243

STS88. Schulz MH, Trischler E, Sarfert TM (1988) SOCRATES: a highly efficient automatic test pattern generation system. IEEE Trans Comput-Aided Des Integr Circuits Syst 7(1):126–137

SHB68. Sellers FF, Hsiao MY, Bearnson LW (1968) Analyzing errors with the Boolean difference. IEEE Trans Comput 17(7):676–683

SSL+92. Sentovich EM, Singh KJ, Lavagno L, Moon C, Murgai R, Saldanha A, Savoj H, Stephan PR, Brayton RK, Sangiovanni-Vincentelli AL (1992) SIS: a system for sequential circuit synthesis. Technical report, University of Berkeley

SPR02. Shao Y, Pomeranz I, Reddy SM (2002) On generating high quality tests for transitions faults. In: Proceedings of the IEEE Asian test symposium, pp 1–8

SP02. Sharma M, Patel JH (2002) Finding a small set of longest testable paths that cover every gate. In: Proceedings of the international test conference, pp 974–982

SS06. Sheini HM, Sakallah KA (2006) Pueblo: a hybrid pseudo-Boolean SAT solver. J Satisfiability, Boolean Model Comput 2(1–4):165–189

SFD+05. Shi J, Fey G, Drechsler R, Glowatz A, Hapke F, Schloeffel J (2005) PASSAT: efficient SAT-based test pattern generation. In: Proceedings of the IEEE annual symposium on VLSI, pp 212–217

Sht01. Shtrichman O (2001) Pruning techniques for the SAT-based bounded model checking problem. In: Proceedings of the correct hardware design and verification methods. Volume 2144 of lecture notes in computer science, pp 58–70

Smi85. Smith GL (1985) Model for delay faults based upon paths. In: Proceedings of the international test conference, pp 342–349

Sne77. Snethen TJ (1977) Simulator-oriented fault test generator. In: Proceedings of the design automation conference, pp 88–93

SSAM93. Srinivasan S, Swaminathan G, Aylor JH, Mercer MR (1993) Algebraic ATPG of combinational circuits using binary decision diagrams. In: Proceedings of the European test conference, pp 240–248

SB91. Stanion T, Bhattacharya D (1991) TSUNAMI: a path oriented scheme for algebraic test generation. In: Proceedings of the international symposium on fault-tolerant computing, pp 36–43

SBS96. Stephan P, Brayton RK, Sangiovanni-Vincentelli AL (1996) Combinational test generation using satisfiability. IEEE Trans Comput-Aided Des Integr Circuits Syst 15(9):1167–1176

SB77. Storey TM, Barry JW (1977) Delay test simulation. In: Proceedings of the design automation conference, pp 492–494

TGA00. Tafertshofer P, Ganz A, Antreich KJ (2000) IGRAINE – an implication graph based engine for fast implication, justification, and propagation. IEEE Trans Comput-Aided Des Integr Circuits Syst 19(8):907–927

TELD09. Tille D, Eggersglüß S, Le HM, Drechsler R (2009) Strutural heuristics for SAT-based ATPG. In: Proceedings of the IFIP/IEEE international conference on very large scale integration

TED10. Tille D, Eggersglüß S, Drechsler R (2010) Incremental solving techniques for SAT-based ATPG. IEEE Trans Comput-Aided Des Integr Circuits Syst 29(7):1125–1130

TK99. Tragoudas S, Karayiannis D (1999) A fast nonenumerative automatic test pattern generator for path delay faults. IEEE Trans Comput-Aided Des Integr Circuits Syst 18(7):1050–1057

Tse68. Tseitin GS (1968) On the complexity of derivation in the propositional calculus. Stud Constr Math Math Logic Part II:115–125

Vel04. Velev MN (2004) Encoding global unobservability for efficient translation to SAT. In: Proceedings of the international conference on theory and applications of satisfiability testing, pp 197–204

WLRI87. Waicukauski JA, Lindbloom E, Rosen BK, Iyengar VS (1987) Transition fault simulation. IEEE Des Test Comput 4(2):32–38

WSGM90. Waicukauski JA, Shupe PA, Giramma DJ, Matin A (1990) ATPG for ultra-large structured designs. In: Proceedings of the international test conference, pp 44–51

WC08. Wang Z, Chakrabarty K (2008) Test-quality/cost optimization using output-deviation-based reordering of test patterns. IEEE Trans Comput-Aided Des Integr Circuits Syst 27(2):352–365

WWW06. Wang L-T, Wu C-W, Wen X (2006) VLSI test principles and architectures. Elsevier

WKS01. Whittemore J, Kim J, Sakallah K (2001) SATIRE: a new incremental satisfiability engine. In: Proceedings of the design automation conference, pp 542–545

WA73. Williams MJY, Angell JB (1973) Enhancing testability of large-scale integrated circuits via test points and additional logic. IEEE Trans Comput 22(1):46–60

WLLH07. Wu C-A, Lin T-H, Lee C-C, Huang C-Y (2007) QuteSAT: a robust circuit-based SAT solver for complex circuit structure. In: Proceedings of design, automation and test in Europe, pp 1313–1318

YCW04. Yang K, Cheng K-T, Wang L-C (2004) Trangen: a SAT-based ATPG for path-oriented transition faults. In: Proceedings of the ASP design automation conference, pp 92–97

YCT08. Yilmaz M, Chakrabarty K, Tehranipoor M (2008) Test-pattern grading and pattern selection for small-delay defects. In: Proceedings of the VLSI test symposium

YCT10. Yilmaz M, Chakrabarty K, Tehranipoor M (2010) Test pattern selection for screening small-delay defects in very-deep submicron integrated circuits. IEEE Trans Comput-Aided Des Integr Circuits Syst 29(5):760–773

ZMMM01. Zhang L, Madigan CF, Moskewicz MW, Malik S (2001) Efficient conflict driven learning in a Boolean satisfiability solver. In: Proceedings of the international conference on computer-aided design, pp 279–285

Index

S. Eggersglüß and R. Drechsler, *High Quality Test Pattern Generation and Boolean Satisfiability*, DOI 10.1007/978-1-4419-9976-4,